Semiconductor Devices in Harsh Conditions

Devices, Circuits, and Systems

Series Editor

Krzysztof Iniewski
Emerging Technologies CMOS Inc.
Vancouver, British Columbia, Canada

PUBLISHED TITLES:

Embedded and Networking Systems:
Design, Software, and Implementation
Gul N. Khan and Krzysztof Iniewski

Energy Harvesting with Functional Materials and Microsystems
Madhu Bhaskaran, Sharath Sriram, and Krzysztof Iniewski

Gallium Nitride (GaN): Physics, Devices, and Technology
Farid Medjdoub

Graphene, Carbon Nanotubes, and Nanostuctures:
Techniques and Applications
James E. Morris and Krzysztof Iniewski

High-Speed Devices and Circuits with THz Applications
Jung Han Choi

High-Speed Photonics Interconnects
Lukas Chrostowski and Krzysztof Iniewski

High Frequency Communication and Sensing:
Traveling-Wave Techniques
Ahmet Tekin and Ahmed Emira

High Performance CMOS Range Imaging:
Device Technology and Systems Considerations
Andreas Süss

Integrated Microsystems: Electronics, Photonics, and Biotechnology
Krzysztof Iniewski

Integrated Power Devices and TCAD Simulation
Yue Fu, Zhanming Li, Wai Tung Ng, and Johnny K.O. Sin

Internet Networks: Wired, Wireless, and Optical Technologies
Krzysztof Iniewski

Introduction to Smart eHealth and eCare Technologies
Sari Merilampi, Krzysztof Iniewski, and Andrew Sirkka

Ionizing Radiation Effects in Electronics: From Memories to Imagers
Marta Bagatin and Simone Gerardin

Labs on Chip: Principles, Design, and Technology
Eugenio Iannone

Laser-Based Optical Detection of Explosives
Paul M. Pellegrino, Ellen L. Holthoff, and Mikella E. Farrell

Low Power Emerging Wireless Technologies
Reza Mahmoudi and Krzysztof Iniewski

Medical Imaging: Technology and Applications
Troy Farncombe and Krzysztof Iniewski

PUBLISHED TITLES:

Metallic Spintronic Devices
Xiaobin Wang

MEMS: Fundamental Technology and Applications
Vikas Choudhary and Krzysztof Iniewski

Micro- and Nanoelectronics: Emerging Device Challenges and Solutions
Tomasz Brozek

Microfluidics and Nanotechnology: Biosensing to the Single Molecule Limit
Eric Lagally

MIMO Power Line Communications: Narrow and Broadband Standards, EMC, and Advanced Processing
Lars Torsten Berger, Andreas Schwager, Pascal Pagani, and Daniel Schneider

Mixed-Signal Circuits
Thomas Noulis

Mobile Point-of-Care Monitors and Diagnostic Device Design
Walter Karlen

Multisensor Attitude Estimation: Fundamental Concepts and Applications
Hassen Fourati and Djamel Eddine Chouaib Belkhiat

Multisensor Data Fusion: From Algorithm and Architecture Design to Applications
Hassen Fourati

MRI: Physics, Image Reconstruction, and Analysis
Angshul Majumdar and Rabab Ward

Nano-Semiconductors: Devices and Technology
Krzysztof Iniewski

Nanoelectronic Device Applications Handbook
James E. Morris and Krzysztof Iniewski

Nanomaterials: A Guide to Fabrication and Applications
Sivashankar Krishnamoorthy

Nanopatterning and Nanoscale Devices for Biological Applications
Šeila Selimović

Nanoplasmonics: Advanced Device Applications
James W. M. Chon and Krzysztof Iniewski

Nanoscale Semiconductor Memories: Technology and Applications
Santosh K. Kurinec and Krzysztof Iniewski

Novel Advances in Microsystems Technologies and Their Applications
Laurent A. Francis and Krzysztof Iniewski

Optical, Acoustic, Magnetic, and Mechanical Sensor Technologies
Krzysztof Iniewski

PUBLISHED TITLES:

Testing for Small-Delay Defects in Nanoscale CMOS Integrated Circuits
Sandeep K. Goel and Krishnendu Chakrabarty

Tunable RF Components and Circuits: Applications in Mobile Handsets
Jeffrey L. Hilbert

VLSI: Circuits for Emerging Applications
Tomasz Wojcicki

Wireless Medical Systems and Algorithms: Design and Applications
Pietro Salvo and Miguel Hernandez-Silveira

Wireless Technologies: Circuits, Systems, and Devices
Krzysztof Iniewski

Wireless Transceiver Circuits: System Perspectives and Design Aspects
Woogeun Rhee

FORTHCOMING TITLES:

Diagnostic Devices with Microfluidics
Francesco Piraino and Šeila Selimović

Magnetic Sensors: Technologies and Applications
Laurent A. Francis, Kirill Poletkin, and Krzysztof Iniewski

Nanoelectronics: Devices, Circuits, and Systems
Nikos Konofaos

Noise Coupling in System-on-Chip
Thomas Noulis

Radio Frequency Integrated Circuit Design
Sebastian Magierowski

X-Ray Diffraction Imaging: Technology and Applications
Joel Greenberg and Krzysztof Iniewski

Semiconductor Devices in Harsh Conditions

Edited by
Kirsten Weide-Zaage ● Malgorzata Chrzanowska-Jeske

Managing Editor
Krzysztof Iniewski

CRC Press
Taylor & Francis Group
Boca Raton London New York

CRC Press is an imprint of the
Taylor & Francis Group, an **informa** business

CRC Press
Taylor & Francis Group
6000 Broken Sound Parkway NW, Suite 300
Boca Raton, FL 33487-2742

First issued in paperback 2020

© 2017 by Taylor & Francis Group, LLC
CRC Press is an imprint of Taylor & Francis Group, an Informa business

No claim to original U.S. Government works

ISBN-13: 978-1-4987-4380-8 (hbk)
ISBN-13: 978-0-367-65636-2 (pbk)

Library of Congress Cataloging-in-Publication Data

Names: Weide-Zaage, Kirsten, editor. | Chrzanowska-Jeske, Malgorzata, editor.
Title: Semiconductor devices in harsh conditions / edited by Kirsten
Weide-Zaage and Malgorzata Chrzanowska-Jeske.
Description: Boca Raton : Taylor & Francis, a CRC title, part of the Taylor &
Francis imprint, a member of the Taylor & Francis Group, the academic
division of T&F Informa, plc, 2017. | Series: Devices, circuits & systems
series
Identifiers: LCCN 2016022751 | ISBN 9781498743808
Subjects: LCSH: Semiconductors--Reliability. | Extreme environments. |
Environmental testing.
Classification: LCC TK7871.85 .S449 2017 | DDC 621.3815/2--dc23
LC record available at https://lccn.loc.gov/2016022751

Visit the Taylor & Francis Web site at
http://www.taylorandfrancis.com

and the CRC Press Web site at
http://www.crcpress.com

Contents

Section I Radiation

Section II Sensors and Operating Conditions

Section III   Packaging and System Design

Foreword

Many applications need electronics that must be specially designed to reliably operate in unique harsh environments. One may ask what makes an environment harsh for electronic device operation? In a simple way we can respond that it is a set of conditions that can cause faulty operation of a device or can destroy completely that device. Some examples of harsh environments for electronics are extreme temperatures and temperature cycles, water and other liquids including human liquids for biomedical devices, high humidity levels, electrostatic discharge (ESD), electromagnetic interference (EMI) and radiation, vibrations, physical impact, and possibly others. Many of these harsh conditions can coexist and make a designer's job even more challenging as there is a lack of data on mixed-model environments.

It is not only semiconductor devices that have to be designed to withstand these harsh conditions but also the packaging systems. Temperature, both extreme cold and extreme heat, is probably the single most challenging condition for packaging solutions especially for applications in the downhole oil and gas industry, space exploration, and automobiles. Additional harsh conditions are usually present, making the job of finding a good solution extremely tough. Further, there is a high cost of developing these solutions but required numbers of such devices are typically low.

Designing electronic systems for harsh conditions is a very broad topic and is not possible to cover in one book, and this is not what the editors of this book had in mind. They have concentrated on three key areas: devices in the presence of radiation, sensors in harsh environments from high temperature to human liquids, and system packaging for extreme environments. The contributing authors are experts from industry and academia from four continents: North America, South America, Europe, and Asia. Readers will find such interesting topics as packaging for harsh environments, biomedical sensors, and radiation-resistant designs.

Having been involved for many years in improving the reliability of devices when submitted to harsh environmental conditions, I am very glad to support the initiative of the editors of this book. Indeed, as electronics are used in all sectors of human activity, and hence, in many cases, in very hard conditions, it is very important to assess lifetime keeping in mind the lowest failure rate. To do that, understanding and modelling of failure mechanisms under high stresses is essential and can be obtained with experimental testing as well as simulation. Such a reliability-oriented design will be necessary to select pertinent technological solutions when facing a given hard mission profile, and at the same time forecast a lifetime corresponding to a maximum acceptable failure rate in a dedicated application.

I am sure that this book, prepared by Dr Kirsten Weide-Zaage and Dr Malgorzata Chrzanowska-Jeske, is an important contribution to the "reliability challenge" faced now by the electronic devices industry. Thank you for this useful initiative!

Professor Yves Danto
University of Bordeaux France

Preface

The unique environmental conditions for microelectronic applications in medicine, transportation, energy and space require design and manufacturing processes that can withstand exposure to moisture, radiation, extreme temperature and pressure and corrosive chemicals. There is intense interest in understanding the behaviour of microelectronic and nanoelectronic components when exposed to such harsh environments, developing designs which can endure rapid environmental changes, and understanding the role of packaging to protect the functional elements of a microsystem in these hostile conditions.

While each category of applications has specific needs and requirements, they do share broad needs which can often be addressed with general solutions.

For medical devices, the obvious focus when introducing microelectronics is the effect on the patient. Devices and the materials used need to be biocompatible: have no toxic effect on the body, not provoke an immune response, and reliably fulfil their function for the lifetime of the device. However, to reliably fulfil their functions, these devices must face a barrage of challenges, from the high temperatures of autoclave sterilisation (high-pressure steam at 121°C) to the moist environment on and in the human body, in saliva, blood, urine and the acids and enzymes of the digestive system.

The temperature requirements can be even more demanding for automotive applications because a growing number of control units and sensor systems are found in the engine compartment in contact with hot lubricants or gasoline. Typical operating temperatures found in engines are much hotter than the maximum operating temperature for most integrated circuits (often specified below 125°C). Similar challenges exist in the rail and aerospace industries.

The energy industry often compounds high temperature and pressure with the corrosive chemicals used to drill and recover hydrocarbons. Deep-sea oil drilling and recent innovations like hydraulic fracturing put great stress on sensors and controllers. Some renewable energy alternatives like concentrated solar power also operate at several hundred degrees Celsius. But perhaps the most extreme conditions in the energy industry are found in nuclear reactors, where in addition to corrosive compounds and high temperature and pressure, designers face the maelstrom of ionising radiation.

The radiation hardness of components is vital even beyond applications where radiation is expected. For example, radiation hardness is essential for semiconductor integrated circuit fabrication processes such as X-ray lithography, plasma etching and reactive ion etching. In medical or automotive components, a single-event effect due to natural radiation can lead, in extreme cases, to a complete failure of the components.

Radiation and other extremes can be a particular challenge in scientific exploration of space, extreme environments like the deep sea and volcanoes, and the very limits of physics (through the use of particle accelerators). When exposed to the harsh radiation and temperature ranges of space, electronic components may degrade or fail due to the effects of ionising radiation and thermal stress. Often, no maintenance is possible on a space probe (the Voyager probes launched in 1977 continue to relay data as they travel beyond the solar system) or maintenance is at great expense (the Hubble telescope requires manned space missions). Even scientific missions on Earth rely on sensors functioning reliably over a known time period.

Therefore, to predict the length of survival of these systems, it is important to carefully study reliability of microelectronic devices in the presence of radiation and other harsh conditions. To prevent material degradation, various self-healing methods can be used to help increase device reliability. In some special cases, a new package design can be developed specifically for devices operating in extreme environmental conditions.

Cost can be a restraining factor; the expense of designing custom components for harsh conditions, particularly radiation hardness, can make the prospect of using commercial off-the-shelf components very attractive. Recently, silicon devices built on insulators and devices made of compound semiconductors from groups III–V (like nitrides – AlN, GaN, and InN) have been developed and tested specifically for applications that include operation under severe conditions.

With increasing device complexity and stronger demand for reliable and cost-effective operation in diverse and extreme environments, the capabilities of standard electronics are approaching their limits. This book presents insights from a group of diverse researchers into a broad set of material, design and manufacturing problems being hotly debated as research continues into microelectronic applications in harsh environments.

The book is divided into three parts. Part I contains three chapters discussing issues related to the operation of electronic devices affected by radiation.

Chapter 1 provides an overview of the challenges posed by radiation to the normal functioning and lifetime duration of electronics deployed on spacecraft, and a case study in which several solutions have been implemented in a space-worthy system. The architecture proposed has been selected as part of the third-generation Meteosat to be launched in 2018. The described system was validated by means of fault injection campaigns using a software-based system. Results gathered in these experiments show that the system is suitable for operation in the presence of space radiation.

Chapter 2 examines the issue of soft errors induced by natural radiation at the ground level in current and future complementary metal oxide semiconductor digital circuits. A brief description of the natural radiation environment at the ground and atmospheric levels is given. The physics and the underlying mechanisms of soft errors at the silicon level are summarised. The main mechanisms of interaction between atmospheric radiation, circuit materials and the electrical response of transistors, cells and complete circuits are depicted. The authors discuss soft error characterisation using accelerated and real-time tests, modelling and numerical simulation issues, as well as the radiation response of advanced technologies.

In Chapter 3, simulations of single-event effects in fully depleted silicon-on-insulator transistors and static random access memory cells are presented and compared with the same effects in transistors made with traditional bulk complementary metal oxide semiconductor technology. Impacts were simulated in different locations on the transistor at different impact angles, whereas previous works considered the impact just at a 0° angle. The comparison was performed using two-dimensional technology computer-aided design simulations.

The three chapters in Part II cover sensors for medical and harsh environmental applications and devices under high-temperature operations.

Electrical biosensors, holding the greatest promise for the early detection of ovarian cancer, are described in Chapter 4. They integrate a sensing element and a signal transducer into one convenient device, decreasing the cost and time required for traditional laboratory testing, as well as improving portability. The authors discuss the application of primary electrical biosensing techniques and specifically electrochemical impedance spectroscopy, cyclic voltammetry and potential step chronoamperometry to the development of these biosensors. A rational approach to the design of a biosensor that is capable of detecting all three relevant biomarkers, CA-125, He4 and CEA, simultaneously, is presented. The described electronic platform biosensors are capable of detecting multiple biomarkers for ovarian cancer within the clinically relevant range and low concentrations of these biomarkers.

Chapter 5 describes the development of sensors and sensor systems for harsh environments. Combustion optimisation and emission control in small-scale boilers is presented as an illustrative application that can benefit by the employment of a multimeasurand sensor system able to cope with harsh environmental conditions. This is one of many various applications, with high commercial impact, that can profit from using systems designed and built to withstand harsh environmental conditions. Considerations regarding circuitry, packaging, system integration and testing are discussed, together with a detailed analysis of each sensing device.

Chapter 6 focuses on nitride (III-N) devices feasible for next-generation harsh environment applications due to their unique material properties in extreme conditions. The author begins with a brief discussion on the material fundamentals of III-N semiconductors, and next presents an introductory review of the development of III-N electronic devices. The chapter concludes with a summary of the radiation effect on III-N materials and devices.

Part III contains four chapters covering packaging processes and reliability enhancement issues for harsh operating environments.

Chapter 7 gives an overview of different packaging techniques enabling electronic circuits to operate in harsh environments. The demand for new and improved technologies is increasing continuously. The main challenge is to protect electronic circuits from an external thermal influence and from fast thermal gradients. Other harsh environmental challenges are radioactive, chemical, electromagnetic and high-pressure surroundings. For each of these environmental conditions, specialised encapsulation for the packaging must be chosen. The authors discuss a ceramic often used to ensure a long-term stability of about 10 years.

In Chapter 8, the mechanism for corrosion of lead-free solders in harsh environments (such as marine conditions or acid rain) is presented. When exposed to humidity, lead-free solders are susceptible to galvanic corrosion, explained in terms of electrochemical migration. For illustration, the authors provide several case studies on the corrosion behaviour of different lead-free solders, and present experimental methods to investigate corrosion. The performance, reliability and functionality of industrial standards for microelectronic packaging for harsh environments are also discussed.

Developments in recent calibration methodology for self-healing circuits and systems are discussed in Chapter 9. These systems provide a means for performance and reliability enhancement in the presence of various degradation mechanisms, from deep submicron effects to harsh operating environments. The self-healing calibration problem is complex and requires robust optimisation strategies. The authors describe direct search optimisation algorithms with which they solve practical self-healing calibration problems. Several test cases demonstrate the effectiveness of their approach.

In Chapter 10, the influence of thermal cycling and electromigration on performance of a copper-filled through silicon via (TSV) is presented. These TSVs are a common component of three-dimensional integration concepts in microelectronics. The interfaces of the resulting Cu/Si composite material are affected by diffusional interfacial sliding, which leads to TSV intrusion or protrusion, and affects transistor performance and the reliability of interconnects close to the TSV. These obstacles may limit the use of stacked integrated circuits under harsh environment conditions. The relevance of diffusional interfacial sliding for TSV motion is shown using a three-dimensional finite-element model.

In this book, we introduce the reader to a number of challenges for the operation of electronic devices in various harsh environmental conditions. While some chapters focus on measuring and understanding the effects of these environments on electronic components, many also propose design solutions, whether in choice of material, innovative structures or strategies for amelioration and repair. Many applications need electronics designed to operate in harsh environments, and we hope that our readers will find, in this collection of topics, tools and ideas useful in their own pursuits and of interest to their intellectual curiosity.

The experience of planning, editing and assembling this book has been rewarding for us both. We have learned how broad and varied this topic is based on the contributions of all our authors, who work in very different areas, all bound by common concerns. We hope the reader will learn at least as much as we have of these fascinating topics. We first wish to acknowledge the hard work of all the authors and thank them for their high-quality contributions in this critical subject area. We would like to thank Marcin Jeske, from Maximum Velocity Consulting LLC, for his generous help in editing and polishing of chapters and for his contribution to this preface. We would also like to thank series editor Krzysztof Iniewski for inviting us to contribute to the 'Devices, Circuits, and Systems' book series of CRC Press, and the editorial staff at CRC Press for their help in assembling this book.

Malgorzata Chrzanowska-Jeske
Kirsten Weide-Zaage

MATLAB® is a registered trademark of The MathWorks, Inc. For product information, please contact:

The MathWorks, Inc.
3 Apple Hill Drive
Natick, MA 01760-2098 USA
Tel: 508 647 7000
Fax: 508-647-7001
E-mail: info@mathworks.com
Web: www.mathworks.com

Editors

Kirsten Weide-Zaage is privatdozent (senior lecturer) in the field of microelectronics on the faculty of electrical engineering and computer science at the Gottfried Wilhelm Leibniz Universität in Hannover, Germany. She studied physics with a focus on biophysics, and received her PhD in electrical engineering in the field of migration effects in interconnects. In 2011, she completed her habilitation in the field of microelectronics on the simulation of failure mechanisms in chips and packaging. Since 1988, she has been working on the faculty of electrical engineering and computer science. From 1991 to 2014, she was with the Information Technology Laboratory as a researcher and leader of the simulation group 'Robust Electronics', working in the field of interconnect and package reliability. In 2015, she moved with her renamed group, 'Reliability: Simulation and Risk Analysis', to the Institute of Microelectronic Systems.

Her main research activities are in the field of thermal-electrical-mechanical static and dynamic simulation of microelectronic reliability. These research activities focus on diffusion processes like migration effects in interconnects, contacts, traces as well as solder, the growth of intermetallic compounds, corrosion and package simulation in terms of optimisation. Another topic of the group is technology computer-aided design simulation of the behaviour of transistor cells in terms of processing, mechanical stress, temperature and radiation.

Dr Weide-Zaage has served in various roles on the technical, steering and organising committees of international and national conferences and workshops. She is author of a book and more than 100 scientific articles, including journal and conference publications, invited papers and book chapters. In 2010, her group achieved a best paper award from the European Symposium on Reliability of Electron Devices.

Malgorzata Chrzanowska-Jeske is professor of electrical and computer engineering, and director of the VLSI & Emerging Technology Design Automation Laboratory at Portland State University, Oregon. She joined the Electrical and Computer Engineering Department at Portland State University in 1989 and was department chair from 2004 to 2010. Previously, she served on the faculty of the Technical University of Warsaw, Poland, and as a design automation specialist at the Research and Production Center of Semiconductor Devices in Warsaw, Poland. She holds an MS degree in electronics engineering from the Technical University of Warsaw and a PhD degree in electrical engineering from Auburn University, Alabama.

Her research interests include computer-aided design for very large-scale integration circuits, Mixed-Signal System-on-Chip (MS-SOCs), three-dimensional integrated circuits, nanotechnology, design for manufacturing, and design issues in emerging technologies. She has presented tutorial, keynote and invited talks at various international conferences and events and published more than 150 technical papers. She serves as a panellist and reviewer for the National Science Foundation, and as a reviewer for the National Research Council Canada and many international journals and conferences. Her research has been supported by the National Science Foundation and industry.

Dr Chrzanowska-Jeske has served in various roles on the technical, steering and organising committees of many international conferences and workshops, and as senior editor, associate editor and guest editor of international journals. From 2008 to 2013, she served two terms on the Board of Governors of the Institute of Electrical and Electronics Engineers (IEEE) Circuits and Systems Society, where she was also the chair of the Distinguished Lecturer Program, the chair and founding member of Women in Circuits and Systems and a representative for Regions 1–7 on the Circuits and Systems Society Board of Governors. Currently, she serves as vice president for technical activities for the IEEE Nanotechnology Council. She received the Best Paper Award from IEEE Alabama Section for the best IEEE Transactions paper in 1990, a 1995 Design Automation Conference Scholarship Award, and from IEEE Council on Electronic Design Automation, the 2008 Donald O. Pederson Best Paper Award in IEEE Transactions on Computer-Aided-Design of Integrated Circuits and Systems.

Contributors

Syed Zeeshan Ali
Cambridge CMOS Sensors Ltd.
Cambridge, United Kingdom

Nicolas André
Université Catholique de Louvain
Louvain-la-Neuve, Belgium

Jean-Luc Autran
Aix-Marseille University
Marseille, France

Walter Calienes Bartra
Universidade Fedral do Rio Grande
    do Sul
Porto Alegre, Brazil

Sebastian Bengsch
Institut fuer
    Mikroproduktionstechnik
Produktionstechnisches Zentrum
Leibniz Universitaet
Hannover, Germany

Octavian Buiu
Honeywell
Bucharest, Romania

Cornel Cobianu
Honeywell
Bucharest, Romania

Andrea De Luca
University of Cambridge
Cambridge, United Kingdom

Indranath Dutta
School of Mechanical & Materials
    Engineering
Washington State University
Pullman, Washington

Stefano Esposito
Dipartimento di Automatica e
    Informatica
Politecnico di Torino
Turin, Italy

Denis Flandre
Université Catholique de Louvain
Louvain-la-Neuve, Belgium

Laurent A. Francis
Université Catholique de Louvain
Louvain-la-Neuve, Belgium

Paul D. Franzon
Department of Electrical and
    Computer Engineering
North Carolina State University
Raleigh, North Carolina

Julian W. Gardner
University of Warwick
Coventry, United Kingdom

Tae-Kyu Lee
Cisco Systems Inc.
San Jose, California

Guoli Li
School of Physics and Electronics
Hunan University
Changsha, Hunan, China

Li Li
Cisco Systems Inc.
San Jose, California

Ming Liu
School of Mechanical Engineering
    and Automation
Fuzhou University
Fujian Province, Fuzhou, China

Lutz Meinshausen
Globalfoundries
Dresden, Germany

Daniela Munteanu
Aix-Marseille University
Marseille, France

Nor Ilyana Muhd Nordin
Department of Mechanical
    Engineering
University of Malaya
Kuala Lumpur, Malaysia

**Guillaume
Pollissard-Quatremère**
Université Catholique de Louvain
Louvain-la-Neuve, Belgium

Zoltan Racz
Durham University
Durham, United Kingdom

Ricardo Reis
Universidade Federal do Rio Grande
    do Sul
Porto Alegre, Brazil

Suhana Mohd Said
Department of Electrical
    Engineering
University of Malaya
Kuala Lumpur, Malaysia

Bogdan C. Serban
Honeywell
Bucharest, Romania

Shyh-Chiang Shen
School of Electrical and Computer
    Engineering
Georgia Institute of Technology
Atlanta, Georgia

Raj Solanki
Department of Physics
Portland State University
Portland, Oregon

Florin Udrea
University of Cambridge,
Cambridge, United Kingdom

Massimo Violante
Dipartimento di Automatica e
    Informatica
Politecnico di Torino
Turin, Italy

Andrei Vladimirescu
Institut Supérieur d'Electronique de
    Paris
Paris, France

A. M. Whited
Oak Ridge National Laboratory
Oak Ridge, Tennessee

Tracy Wotherspoon
Microsemi
Caldicot, United Kingdom

Marc Christopher Wurz
Institut fuer
    Mikroproduktionstechnik
Produktionstechnisches Zentrum
Leibniz Universitaet
Hannover, Germany

Eric J. Wyers
Department of Engineering and
    Computer Science
Tarleton State University
Stephenville, Texas

Yun Zeng
School of Physics and Electronics
Hunan University
Changsha, Hunan, China

Section I

Radiation

1

Commercial Off-the-Shelf Components in Space Applications

Stefano Esposito and Massimo Violante

CONTENTS

ABSTRACT Space applications have special requirements due to the particular environment into which they are deployed. Space is a particularly harsh environment for electronic components, because it poses several threats to normal functioning and lifetime duration of electronics deployed in spacecraft. In this chapter, we offer an overview of the problems introduced by radiation and of the solutions developed so far. Then, we present a case study in which several such solutions have been implemented in a space-worthy system. We present the architecture proposed in [30], which was developed in the framework of the ESA HiRel Programme and has been selected as part of the third-generation Meteosat to be launched in 2018. The system described so far was validated by means of fault injection campaigns. A software-based fault injection system was used. Results gathered in these experiments show that the system is suitable for operations in a space radiation environment.

1.1 Introduction

Space applications have always needed to cope with the particular environment into which they are deployed. Space is a particularly harsh environment for electronic components, because it poses several threats to normal functioning and lifetime duration of electronics deployed in spacecraft. One of the most challenging problems in a space application is the presence of high-energy particles interfering with the systems [5]. Radiation is naturally present all around the earth, coming from several sources, like the sun or outer space, and from the interaction among these particles and the earth's magnetic field. All this radiation interferes in several ways with the electronic components and may cause severe malfunctioning (like single-event upsets [SEUs], which consist of random changes of the content of memory bits, or single-event latch-ups, which consist of the destruction of the component due to the activation of parasitic structures leading to short circuits), which could in turn threaten the mission success.

One of the most critical components of an electronic system, in this aspect, is the central processing unit (CPU). Its vulnerability is due mostly to the high level of integration achieved in the manufacturing of CPUs: more advanced manufacturing technologies are also more subject to damage and misbehaviour due to radiation effects [1]. In time, several solutions have been developed to harden CPUs and other key components, like field-programmable gate arrays (FPGAs) and application-specific integrated circuits (ASICs), against radiation's effects [2–4]. Special components designed specifically to resist radiation, called radiation-hardened (RadHard) components, have been developed for space applications. Although these components can operate correctly even in a radiation environment, their main flaw is that the resilience to radiation is paid mostly by a reduction of performance, both for direct effects of the solutions implemented to resist radiation and for effects of the special process of design, manufacturing and certification that RadHard components must go through. This process is usually longer than the design cycle of commercial components, resulting in RadHard components lagging behind the technology cutting edge [2].

Some of the basic functions of a spacecraft, like flight control and housekeeping, are not very demanding in performance terms, but payload applications' performance requirements are quickly growing beyond the capability of the RadHard components. This is why the space industry has been looking for and adopting commercial off-the-shelf (COTS) components for use in space applications, introducing several new solutions to achieve a radiation resilience at least comparable to that achieved by RadHard components.

In this chapter, we offer an overview of the problems introduced by radiation and of the solutions developed so far. Then, we present a case study in which several such solutions have been implemented in a space-worthy system.

The chapter is organised as follows. In Section 1.2, a background is provided with a focus on radiation effects and the space radiation environment. Section 1.3 presents redundancy solutions. Section 1.4 describes the case study and provides a general description of how evaluation of a COTS-based system can be performed. Finally, Section 1.5 presents some conclusions.

1.2 Background

This section presents key concepts and background. First, we define some of the terminology used in the rest of the chapter, and then we explore how radiation affects the electronic circuits, and how these can be modelled. The contents of this section can be found in literature from various sources; we refer primarily to [6].

1.2.1 Terminology

In this subsection, the main terms used in the continuation of the chapter are defined.

All electronic circuits are subject to different phenomena that may cause misbehaviour of the circuit. All these phenomena cause in the circuit some kind of *defect*, that is, a difference between the expected circuit and the actual one. Possible defects include the structural damage due to wear-out, manufacturing errors and unexpected delays. Each defect causes the system to behave in a way different with respect to the expected one. Differences in behaviour of the circuit are called *errors*. An error is the effect of a defect on the behaviour of the circuit. Errors can be classified as suggested in [6]:

- *Silent*: The error had no effect on the outputs of the system.
- *Failure*: The error caused a difference in the output of the system with respect to the expected output without any warning to the user.
- *Detected*: The error caused a difference in the output of the system with respect to the expected output and a warning has been issued to the user.
- *Timeout*: The error caused an unexpected delay in the production of the output. Even though the output may be correct, the fact that it

was produced with a delay reduces or cancels its utility. The additional delay may be unacceptable for the user application, possibly with catastrophic consequences.

Since it would not be possible to analyse each and every defect that might affect a given circuit, test engineers usually refer to *faults*. A fault is the model of a defect at a given abstraction level. There are several *fault models*, each capturing a given aspect of a class of defects. Many fault models have been developed and used; a few are described below, as they are the most pertinent to the scope of this chapter.

First, faults can be classified as follows:

- *Persistent or permanent*: A fault that stays in the circuit after the cause has been removed. Usually, this kind of fault either is structural, for example, physical damage to the circuit, or may have altered the contents of a read-only memory. A persistent fault could be removed from the system by some action; that is, if a fault interested the memory containing the code memory area, reloading the code might cancel the fault and recover normal system behaviour.

- *Temporary faults*: Faults that have a temporary effect on the circuit. They usually disappear spontaneously from the circuit shortly after their cause has been removed.

The models used for permanent and temporary faults are different. The most used model for permanent faults in complementary metal oxide semiconductor (CMOS) circuits is the *stuck-at* fault. The stuck-at fault model applies to the gate-level description of an electronic component and models all defects that cause a given wire to carry either a logic 1 (stuck-at-1) or a logic 0 (stuck-at-0), regardless of the output of the driving gate. Several defects can be modelled by stuck-at. For instance, damage to the gate can cause it to never connect the pull-up network; thus, its output would always be 0 regardless of the inputs. Another model is the *delay* model. The delay fault model applies to the gate-level description of an electronic component and models all defects that do not alter the output of a net, but its timing. A wrong timing could cause the network to violate timing constraints of the memory element latching its output, thus changing the behaviour of the circuit over time.

When considering radiation effects, a designer mostly faces temporary faults. Radiation can interfere with the behaviour of a circuit in many ways. When a silicon-based integrated circuit is considered, many effects are due to the interaction between the particles and the silicon lattice of the substrate. There are two main mechanisms of interaction:

- *Direct*: A charged particle can directly interact with the silicon lattice, inducing currents by displacing electron–hole pairs.

- *Indirect*: A noncharged particle can interact with atoms in the circuits, producing secondary particles by several nuclear mechanisms. The secondary particles include charged particles, which causes currents by displacing electron–hole pairs.

The induced currents are responsible for the changed behaviour of the circuit. A current in a gate can cause its output to change temporarily, whereas a current in a memory cell can change the content of the memory cell. The effects of a single-particle strike on the circuit are called *single-event effects* (SEEs). SEEs include several fault models:

- *Single-event transient (SET)*: The temporary change of the output of a gate in the circuit due to the radiation effect. A SET can be described as an unexpected glitch in the circuit, one that may propagate as an error if it causes a wrong value to be latched in a flip-flop.
- *Single-event upset*: The change of content of a memory cell in the circuit due to the radiation effect. The memory cell is still working properly; thus, a subsequent write operation will restore the correctness of its content. However, this kind of fault can propagate as an error if the wrong content is used in computation before it can be corrected.
- *Single-event functional interruption (SEFI)*: The radiation affects the system in a way that its function is compromised and it cannot be recovered except by a reset operation. The circuit is not damaged, and if the system is brought again in a safe state, it can continue operations; however, it is not capable of doing so autonomously.

When a system is subject to radiation for a long period of time, it can accelerate the wear-out of the circuits. This is due to the physical damage particles can inflict on the circuits. Although this kind of effect is out of the scope of this chapter, it is worth noticing that long-term effects of radiation are usually analysed as *total ionising dose* (TID) effects and must be accounted for when selecting a component for space applications, to ensure the component can endure the space environment in which it will be deployed for the mission duration, without being destroyed.

1.3 Error Detection and Fault Tolerance

In this chapter, the focus is on error affecting the CPU of a processor-based system. This section offers a brief survey of the major techniques for error detection and fault tolerance.

To detect errors, one must first select either persistent or transient faults as target of detection.

The processor itself cannot detect persistent faults. As such, this kind of fault must be detected by an external subsystem, as a *watchdog* [7,8]. A watchdog is an external hardware, monitoring the activity of the processor. If the processor activity is not as expected, the watchdog detects the error, enabling a recovery action. There are two main kinds of watchdogs:

1. *Watchdog timers (WDTs)*: Simple timers that the CPU must reset periodically. If the processor fails to reset a WDT before it expires, an error is detected. WDTs can be simple or windowed. A simple WDT measures a time interval, within which the CPU must reset the WDT. A windowed WDT has two thresholds: a maximum and a minimum duration. The CPU must reset a windowed WDT within the maximum threshold, but not before the minimum threshold.

2. *Watchdog processors (WDPs)*: More complex components than the WDTs and can detect more than just persistent faults. A WDP can implement all the functionalities of a WDT, besides other functionalities that help in identifying problems in the control flow of the software. Mostly, WDPs are used to reduce the overhead of control-flow check (CFC) techniques.

Transient faults can be detected in several ways, using both hardware and software approaches or a combination of both.

1.3.1 Hardware-Based Solutions for Error Detection and Fault Tolerance

Hardware approaches entail the duplication or triplication of hardware, which means duplication or triplication of a processor, and possibly of its companion chips, in the scope of this chapter. The *SCS750 architecture* [9] is an example of a hardware-based solution still used. In this architecture, the processor module is triplicated and the outputs are compared to perform a majority vote. Other solutions are also implemented to improve system reliability, such as a memory protected by an error-correcting code (ECC), and a scrubber, which periodically outputs all processor contents to the error-corrected memory and resynchronises all processors in the systems, in order to ensure that processors affected by a transient fault are effectively recovered.

1.3.2 Software-Based Solutions for Error Detection and Fault Tolerance

Software-based solutions describe how to modify the software in order to obtain error detection. They can be classified as data-hardening solutions or CFC solutions, depending on whether they are designed to protect the data used by the CPU for computation or focus on ensuring that at any time, the correct execution path is being followed.

1.3.2.1 Data-Hardening Solutions

Data-hardening solutions use the repetition of computation in the CPU, in order to compare results and detect errors. Different redundancy granularities can be used in software-based solutions.

Instruction duplication is the finest-granularity approach, and has been proposed in many solutions [10–14]. The main idea of instruction duplication is to repeat each instruction twice and compare the data in order to detect errors. In instruction-level duplication solutions, each variable is duplicated and the replicas are compared for agreement after each read operation. A mismatch in variable replicas means an error has been detected. Instruction-level duplication can be applied at either the assembly level or the high level.

Procedure-level duplication is a purely software technique with a coarser granularity of duplication with respect to instruction-level duplication [15]. In this solution, some procedures would be modified to implement instruction-level duplications, whereas others would be left unmodified. Modified procedures would be able to detect errors, whereas unmodified procedures would be called twice to detect errors.

The coarse granularity of duplication is the *program-level duplication*. In this approach, an entire program is duplicated and executed twice, implementing a temporal redundancy in a *virtual duplex system* (VDS) [16], as shown in Figure 1.1.

Although this approach introduces a higher error latency than both instruction- and procedure-level duplication, it can benefit from parallelisation to reduce performance overhead [17]. It can also benefit from *N-versioning* to reduce common mode errors [16,18–20].

The *Proton100k* computer [21] uses a software solution applicable to very long instruction word (VLIW) processors to implement a schema known as *temporal triple modular redundancy* (TTMR). In TTMR, the software executes twice and results are compared to detect errors. If an error is detected, a

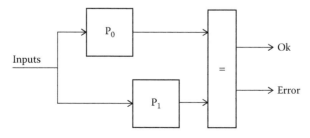

FIGURE 1.1
VDS configuration.

third execution is performed and the correct output is selected through a majority vote.

1.3.2.2 Control-Flow Check Solutions

Another set of software-based solutions is based on the attempt to preserve the program control flow in order to avoid silent data corruption and timing errors which could not be detected by a data-hardening solution, as the ones described above, and CFC techniques are needed. The basic concept of control-flow analysis is the *basic block* (BB). A BB is a portion of code that does not contain jumps, except the last instruction, and is not the target of a jump, except the first instruction. A BB represents a flow of instructions that are executed in sequence without jumps. A program can be described as an oriented graph, whose nodes are the BBs composing the program and whose edges are the jumps connecting the BBs. Such a graph is called a *control-flow graph* (CFG). A CFG allows us to define a relationship among BBs. A BB b_i is a predecessor of a BB b_j if and only if the branch v_{ij} going from b_i to b_j is contained in the program's CFG. Vice versa, a BB b_i is a successor of a BB b_j if and only if the branch v_{ji} going from b_j to b_i is contained in the program's CFG. Describing a program in terms of its CFG allows defining a classification for branches too. As suggested in [6], branches can be

- *Legal*: If they are contained in the CFG
- *Wrong*: If they are contained in the CFG but are taken at an unexpected time, for example, before the predecessor BB execution was completed
- *Illegal*: If they are not contained in the CFG

All but legal branches are *control-flow errors* (CFEs). Several solutions have been developed to detect CFEs. In this subsection, we present an overview of some of the main ones.

Path identification [22] uses a partitioning of the CFG to group BBs. A prime number identified is assigned to each BB in the CFG, and the CFG is partitioned into loop-free intervals. At the end of each loop-free interval, a check is performed to ensure that the execution followed the expected sequence of BBs. During the execution of a loop-free interval, a path identifier *RPI* is computed by multiplying the prime identifiers of each traversed BB. The RPI is an index in a path table containing a path predicate and a next-interval identifier, *NIID*. At the beginning of a loop-free interval, the table is accessed and the RPI is used to retrieve the correct row. An error is detected if the path predicate evaluates false or if the NIID does not correspond to the current interval identified, *CIID*. If all checks pass, the RPI is reinitialised to 1 and computation continues.

Enhanced control-flow checking using assertion (ECCA) [23] uses two specifically designed assertions in order to detect CFEs. This technique uses an *id* variable, which is updated at runtime at the beginning and end of each BB.

A unique prime identifier different from 2 is assigned to each BB. At the beginning of a BB, the id variable is assigned as follows:

$$id = \frac{BID}{(id\%BID)(id\%2)}$$

This formula is called a *SET* assertion. At the end of each BB, a *TEST* assertion is executed, as follows:

$$id = NEXT + \overline{\overline{(id - BID)}}$$

where $NEXT = \prod BID_{\text{successors}}$. By definition, if either $\overline{id\%BID} = 0$, meaning the current BB is not in the successors of the BB in which the id was last updated, or $id\%2 = 0$, meaning that in the last TEST assertion the term $\overline{(id - BID)}$ was 1, the SET assertion would raise a divide-by-zero exception, thus detecting the error.

Yet another control-flow check using assertion (YACCA) [24,25] uses a more complex assertion scheme than ECCA to achieve better performances. To each BB, v_i are assigned two identifiers: $I1_i$, associated with the beginning, and $I2_i$, associated with the end. The TEST assertion updates a *code* variable so that $code = I1_i$ at the beginning of a BB, and at the same time verifies that the *code* variable is equal to the end identifier of a predecessor. The SET assertion updates the same *code* variable so that at the end of the same BB, $code = I2_i$, checking at the same time that the *code* variable is equal to the begin identifier of the same BB. The *code* variable is updated as follows in both assertions:

$$code = (code \,\&\, M1) \oplus M2$$

M1 is a compile time constant depending on the predecessors of the current BB; *M2* is a different compile time constant that depends on both the expected value for *code* and the predecessors of the current BB. The TEST and SET assertions differ by the definition of M1 and M2. These are the definitions of M1 and M2 for the TEST assertion:

$$M1 = \overline{\left(\underset{j:v_j \in pred(v_i)}{\&} I2_j \right) \oplus \left(\underset{j:v_j \in pred(v_i)}{\vee} I2_j \right)}$$

$$M2 = (I2_j \,\&\, M1) \oplus I1_i$$

whereas these are the definitions of M1 and M2 for the SET assertion:

$$M1 = 1$$

$$M2 = I1_i \oplus I2_i$$

1.3.2.3 Fault Tolerance

The solutions described so far only deal with error detection, but a recovery mechanism must be implemented in order to provide proper fault tolerance. In some limited cases, an algorithm can be modified using its mathematical properties to obtain a fault-tolerant algorithm, which includes both error detection and correction, in a technique known as *algorithm-based fault tolerance* (ABFT) [26,27].

More general approaches are based on a detection phase that could be implemented by one of the techniques described in this section, and on a correction phase. To maximise detection and correction, it is critical that the replicas of a program are not affected by common mode errors. To ensure this, the principle of *design diversity* [19] can be used. In design diversity approaches, the replicas of software used in a redundancy scheme are developed differently: different algorithms to implement the same functionalities, different ways of encoding the algorithm (e.g. by using different languages), different compilers and possibly different programmers and teams altogether. N-versioning [20] is a design diversity approach in which *N* versions of the same software run in the system at the same time. At given intervals, each replica saves a state vector. All the vectors from the different instances are compared for consensus. Another approach, called *recovery block* [28], uses *N* variants of a program, but only one is running in the system. At predefined checkpoints, the active variant stores a safe context. A decider performs an acceptance test on the outputs of the active variant. If a test fails, the safe context is restored in a different variant, which then resumes execution.

1.3.3 Hybrid and System-Level Solutions

Combining hardware approaches with software-based approaches, a number of hybrid solutions can be devised. Most such solutions require a companion chip implementing some fault tolerance–specific hardware, such as WDPs. In the following section, a case study implementing a hybrid solution is presented, which was developed by combining different solutions to obtain a high-performance COTS-based space computer. In this subsection, two precursor architectures are presented.

The *duplex multiplexed in time* (DMT) and *dual duplex tolerant to transients* (DT2) architectures [29] are both based on the VDS schema. In DMT, the execution is duplicated in time on the same single microprocessor, while in DT2, the processor is duplicated. Both DMT and DT2 need a companion chip and an ECC-protected memory, both duplicated along with the CPU in the DT2 architecture. Both DMT and DT2 are based on safe-context storage for fault recovery and need special functions to be implemented in the companion chip, which must be RadHard. From the software point of view, DMT and DT2 require that the software is split in three phases: input acquisition, processing and output presentation. Note that the DT2 architecture does not require the duplicated processors to operate in lock-step mode. When an error is detected, both processors receive an interrupt and implement a recovery action.

1.4 Case Study

In this section, we present the architecture proposed in [30], which was developed in the framework of the ESA HiRel Programme and has been selected as part of the third-generation Meteosat to be launched in 2018.

The HiRel architecture includes elements from the solutions described in the previous section, combining them to create a highly reliable, high-performance computer for space applications.

1.4.1 System Overview

A simplified schema of the system is presented in Figure 1.2. The architecture is composed of COTS components implementing all the main system functionalities.

The CPU is a commercial PowerPC (PPC) processor running at a top speed of 1 GHz, coupled with a 512 MiB DDR-II DRAM memory. In the middle, there are two companion chips. One companion chip is implemented in a commercial high-frequency FPGA and is called Bridge-FPGA. The Bridge-FPGA implements some key features for system operations. First, it implements

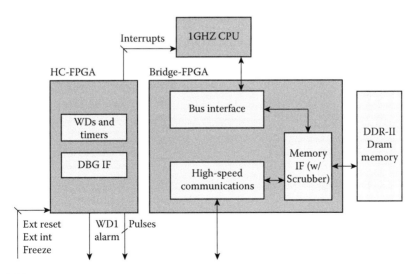

FIGURE 1.2
Simplified system architecture.

a bus interface to the CPU, allowing access from the CPU to all memory-mapped interfaces and peripherals, including the main memory. The memory controller is also implemented in the Bridge-FPGA. Besides the normal memory controller functions, this memory controller also implements a scrubber and an ECC coder/decoder. Finally, the Bridge-FPGA also implements all high-speed communications with the external world, more specifically, a SpaceWire (SpW) controller and a High-Speed Serial Link (HSSL) controller. All high-speed interfaces are connected to a direct memory access (DMA) controller, also implemented in the Bridge-FPGA, allowing data transfer to be performed without the direct intervention of the CPU. At last, the Bridge-FPGA is also responsible for interfacing the CPU with the second companion chip, the Health-Care FPGA (HC-FPGA). HC-FPGA is deployed in a lower-speed FPGA, compared with the FPGA used for Bridge-FPGA. HC-FPGA implements some critical features for the system fault tolerance strategy and for the fault injection system implemented to validate the system:

- *WDT and WDP*: HC-FPGA implements a windowed WDT and a WDP. The WDP implements a signature-based CFC technique, which is described in section 1.4.2. When the WDT is triggered, it begins a panic reaction, stopping the clock and raising a signal on the external interface. The only way to recover from this state is to use either a power cycle or an external reset signal to be managed by a platform computer. The rationale is that an error detected by the WDT is probably an SEFI and as such must be recovered through a system reset.

- *Debug interface*: This interface is used to synchronise the fault injection system with the execution of the software.

- *External interrupt controller*: This controller receives and manages interrupts from the external interfaces and sends them to the CPU.

The HC-FPGA also implements several other low-speed features, such as a Universal Asynchronous Receiver/Transmitter (UART) interface. A critical function of the HC-FPGA is the configuration memory scrubber for the Bridge-FPGA configuration memory. This scrubber grants that faults do not accumulate in the configuration memory of the Bridge-FPGA.

1.4.2 Fault Tolerance Strategy

The system overviewed above was designed to operate as a payload computer on a spacecraft. To ensure the proper functioning in the space radiation environment, a proper fault tolerance solution must be implemented. The solution designed for the system is based on solutions described in the previous sections and is a hybrid one; that is, it uses both software and hardware to achieve detection and tolerance of SEEs.

The solution can be described as a superposition of different layers of fault tolerance. A first layer is composed of a combination of purely software data-hardening techniques. The second is composed of special hardware properly configured and activated by the software at the proper time. To implement the first layer, the software is modified to implement procedure call duplication. The software in the system is partitioned into three phases:

1. *Acquisition (A)*: The inputs are read from communication peripheral buffers.
2. *Processing (P)*: The inputs are processed to generate the outputs.
3. *Presentation (O)*: The outputs are prepared for transmission to the user; that is, the output data are copied onto communication peripheral buffers.

Each phase is implemented as a procedure called twice. Each procedure uses different memory area to store results used by a subsequent phase and to read data passed from the previous phase; thus, data duplication is implemented. Error detection in this layer is achieved by data comparison. The input procedure computes a signature of the input data. The two signatures produced by the two subsequent calls are compared for agreement; if they do not match, an error is detected, triggering error recovery. The processing procedure computes a signature on the output data. As in the previous case, the two signatures produced by the two subsequent calls are compared, and if they do not match, an error is detected, triggering error recovery. Error recovery can be implemented in two ways:

1. *Backward recovery*: When an error is detected in any point in time, execution is reverted back from the start of the input phase.
2. *Forward recovery*: When an error is detected at the end of the two executions, a third execution is performed and the outputs are compared with the first two executions to select the correct one by a majority vote.

The backward recovery introduces a larger overhead when a fault is detected, since it requires both executions to be repeated, but it does not introduce branches that might be the target of a fault, triggering the recovery mechanism when not necessary. On the other hand, the forward recovery is more robust against faults persisting in the system, since the backward recovery actually implements a cyclic behaviour, which may be caught in an endless loop if the error is not effectively removed from the system by the repetition of computation.

The hardware layer for error detection uses features implemented in both the Bridge-FPGA and the HC-FPGA. At the beginning of each phase, the memory protection unit (MPU) implemented in the Bridge-FPGA is used to grant isolation for the data used in all other phases. The MPU allows defining forbidden memory regions in the DDR-II memory. Any access to forbidden regions triggers a hardware exception, managed by an interrupt service routine (ISR) that can flag the current execution as tainted, triggering error recovery at the next checkpoint. This strategy adds a performance overhead, because before switching forbidden regions, the cache must be flushed to avoid an error being triggered by the cache write-back mechanism. This overhead can be reduced by carefully selecting when forbidden regions should be changed. The software can configure the MPU by dedicated application programming interfaces (APIs).

The most important hardware features used by the fault tolerance mechanism are the WDT and the WDP implemented in the HC-FPGA.

The WDT is used to detect SEFI. It is a windowed WDT; thus, it is able to check that computation is completed within a maximum amount of time and not before a minimum amount of time. The two times should be decided through application profiling; the maximum should be the worst-case execution time (WCET) plus some tolerance for the communication delays between the CPU and the WDT registers, whereas the minimum should be the best-case execution time. The software can configure and arm the WDT by using dedicated APIs.

The WDP is used to detect CFEs caused by SEEs. The strategy for CFC is based on the partitioning into blocks of the CFG. A unique signature is assigned to each block at compile time. The WDP is configured with the exact sequence of expected signatures. Moreover, the timing at which each signature should be received can be communicated to the WDP at configuration time. The timing can be determined by application profiling. The software configures and enables the WDP at start-up. At the end of each block, the software it sends the corresponding signature to the WDP. All operations on the WDP are performed through a dedicated API. The WDP detects an error if

- The received signature is not in the set of expected signatures
- The received signature is not the currently expected one
- The signature is not received at the proper time, that is, within a given timeout

When the WDP detects an error, it sends an interrupt to the CPU. The associated ISR can flag the current execution as tainted, triggering the error recovery at the next checkpoint.

1.4.3 Validation

The system described so far was validated by means of fault injection campaigns. A software-based fault injection system was used. A specific ISR was used for fault injection. The ISR was connected to an external interrupt triggered by an external board used for properly timing the injection, labeled *supervisor*. The supervisor was also in charge of stopping the watchdogs using the dedicated freeze signal on the external interface of the target. A workstation was connected to the target system through a Joint Test Action Group (JTAG) interface. The workstation was responsible for loading the software in the target at the beginning of each fault injection experiment, modifying the fault-specific parts of the injection ISR and communicating to the supervisor the time of injection. At the end of each fault injection experiment, the workstation would download the results from target system's memory and compare them with a golden output obtained from a fault-free execution.

Faults were injected into the CPU register file and cache memories and into the Bridge-FPGA configuration memory, because they are the most critical parts of the system and because the robust ECC used to protect memory, and the fact that the HC-FPGA can also be implemented using a RadHard or radiation-tolerant (RadTol) component, makes it superfluous to characterise their behaviour with respect to SEEs.

Injected faults were classified with respect to the outcome of the affected execution, with each execution being isolated from the previous one through a system reset to ensure no fault accumulation could happen. The following classification was used for faults injected in the CPU register file:

- *Silent*: A fault that had no detectable effect on the outputs, or a fault that was detected and recovered successfully.

- *WDT triggered*: A fault that led the system in a state in which it was not able to rearm the WDT, thus triggering its panic action.

- *Signature collision*: The method used to compute the signatures was rather simple to reduce its performance overhead and, as such, subjected to aliasing. In this class are categorised faults that caused the signatures of the outputs to be identical, even though they were different. Using a stronger signature can cancel this class of faults.

- *Illegal instruction*: In this class are the faults that caused an invalid instruction to be fetched. In a live system, these faults would be detected through the hardware exception mechanism; however, the debugger used to connect the workstation with the target system used this exception to implement breakpoints, thus rendering it unusable while in debug mode. This outcome can be considered recovered, even though it has been reported separately.

- *Hardware exception*: The fault triggered a hardware exception. This class is of interest only for the baseline campaign performed without any recovery mechanism in place.

The results gathered are shown in Table 1.1. They show that a significant number of faults, around 18%, would result in a failure in a system without any tolerance mechanism in place (first column). In both recovery scenarios,

TABLE 1.1

Fault Injection on CPU Registers

Classification	None	Recovery Backward	Forward
Silent	717	1790	1843
WDT triggered	—	165	95
Signature collision	—	42	34
Illegal instruction	—	3	28
Hardware exception	102	—	—
Failure	181	—	—
Injected	1000	2000	2000

the failures disappear, although a certain number of faults still lead to WDT to be triggered. The forward recovery shows a better behaviour from this point of view, due to the fact that it cannot be caught in an endless loop, as happens to the backward recovery.

We also injected 2250 faults in cache memories, obtaining no failure at all, thanks to the ECC protecting these memories. The Bridge-FPGA configuration memory was injected with 1000 faults. In backward recovery, two faults triggered the WDT and three faults determined a transmission error, while in forward recovery, one fault triggered the WDT and five faults determined a transmission error.

1.5 Conclusions

COTS-based systems are an interesting perspective for space applications, but such components are not compatible with the characteristics of the space radiation environment. To obtain systems reliable enough to be used in space, special actions are in order. The main way to ensure functionality of a COTS-based system in the space radiation environment is to use some level of redundancy, either software or hardware. Software-implemented hardware fault tolerance (SIHFT) is most promising for COTS components that cannot be modified and whose replication could raise significantly the power budget of the system. Several SIHFT strategies have been developed to obtain a reliable enough system. This chapter presented a case study in which a novel SIHFT strategy based on the integration of several solutions proposed in literature was applied to a new system. The system was then validated through fault injection experiments. Results gathered in these experiments show that the system is suitable for operations in a space radiation environment, and the system has been selected by the European Space Agency to fly on the third generation of Meteosat.

References

1. R. C. Baumann, Radiation-induced soft errors in advanced semiconductor technologies, *IEEE Trans. Device Mater. Reliab.*, vol. 5, no. 3, pp. 305–315, 2005.
2. R. C. Lacoe, J. V. Osborn, R. Koga, S. Brown, and D. C. Mayer, Application of hardness-by-design methodology to radiation-tolerant ASIC technologies, *IEEE Trans. Nucl. Sci.*, vol. 47, no. 6, pp. 2334–2341, 2000.
3. J. J. Wang, Radiation effects in FPGAs, *9th Work. Electron. LHC*, vol. 900, p. 2, 2003.
4. A. Manuzzato, S. Gerardin, A. Paccagnella, L. Sterpone, and M. Violante, Effectiveness of TMR-based techniques to mitigate alpha-induced SEU accumulation in commercial SRAM-based FPGAs, *IEEE Trans. Nucl. Sci.*, vol. 55, no. 4, pp. 1968–1973, 2008.
5. J. L. Barth, C. S. Dyer, and E. G. Stassinopoulos, Space, atmospheric, and terrestrial radiation environments, *IEEE Trans. Nucl. Sci.*, vol. 50, no. 3, pp. 466–482, 2003.
6. O. Goloubeva, M. Rebaudengo, M. Sonza Reorda, and M. Violante, *Software-Implemented Hardware Fault Tolerance*, Springer Science and Business Media, Berlin, 2006.
7. A. Mahmood and E. J. McCluskey, Concurrent error detection using watchdog processors—A survey, *IEEE Trans. Comput.*, vol. 37, no. 2, pp. 160–174, 1988.
8. M. Namjoo and E. J. McCluskey, Watchdog processors and capability checking, in *Twenty-Fifth International Symposium on Fault-Tolerant Computing: "Highlights from Twenty-Five Years,"* Pasadena, CA, June 27–30, 1995, vol. III, pp. 245–248.
9. R. Hillman, G. Swift, P. Layton, M. Conrad, C. Thibodeau, and F. Irom, Space processor radiation mitigation and validation techniques for an 1,800 MIPS processor board, Special Publication ESA SP, in *RADECS—European Conference on Radiation and Its Effects on Components and Systems*, Nooredwijk, the Netherlands, September 15ndse 2003, vol. 2003, no. 536, pp. 347–352.
10. M. Rebaudengo, M. S. Reorda, M. Torchiano, and M. Violante, Soft-error detection through software fault-tolerance techniques, in *Proceedings of 1999 IEEE International Symposium on Defect and Fault Tolerance in VLSI Systems*, Albuquerque, NM, November 1 so 1999, pp. 210–218.
11. M. Rebaudengo, M. Sonza Reorda, M. Torchiano, and M. Violante, A source-to-source compiler for generating dependable software, in *Proceedings of the First IEEE International Workshop on Source Code Analysis and Manipulation*, Florence, Italy, November 10, 2001, pp. 33–42.
12. P. Cheynet, B. Nicolescu, R.Velazco, M. Rebaudengo, M. Sonza Reorda, and M. Violante, Experimentally evaluating an automatic approach for generating safety-critical software with respect to transient errors, *IEEE Trans. Nucl. Sci.*, vol. 47, no. 6, pp. 2231–2236, 2000.
13. N. Oh, P. P. Shirvani, and E. J. McCluskey, Error detection by duplicated instructions in super-scalar processors, *IEEE Trans. Reliab.*, vol. 51, no. 1, pp. 63–75, 2002.
14. G. A. Reis, J. Chang, N. Vachharajani, R. Rangan, and D. I. August, SWIFT: Software implemented fault tolerance, in *International Symposium on Code Generation and Optimization*, San Jose, CA, March 20–23, 2005, vol. 2005, pp. 243–254.

15. N. Oh and E. J. McCluskey, Error detection by selective procedure call duplication for low energy consumption, *IEEE Trans. Reliab.*, vol. 51, no. 4, pp. 392–402, 2002.

16. K. Echtle, B. Hinz, and T. Nikolov, On hardware fault detection by diverse software, in *Proceedings of the 13th International Conference on Fault-Tolerant Systems and Diagnostics*, Bulgaria 1990, pp. 362–367.

17. S. K. Reinhardt and S. S. Mukherjee, Transient fault detection via simultaneous multithreading, in *Proceedings of the 27th International Symposium on Computer Architecture*, Vancouver, BC, June 14, 2000, pp. 25–36.

18. H. Engel, Data flow transformations to detect results which are corrupted by hardware faults, in *Proceedings of the High-Assurance Systems Engineering Workshop*, Niagara, Ontario, October 22, 1996, pp. 279–285.

19. A. Avizienis and J.-C. Laprie, Dependable computing: From concepts to design diversity, *Proc. IEEE*, vol. 74, no. 5, pp. 629–638, 1986.

20. A. Avizienis, The n-version approach to fault-tolerant software, *IEEE Trans. Softw. Eng.*, vol. 11, no. 12, pp. 1491–1501, 1985.

21. D. Czajkowski and M. McCartha, Ultra low-power space computer leveraging embedded seu mitigation, in *2003 IEEE Aerospace Conference Proceedings*, Big Sky, MT, March 8–15, 2003, pp. 5_2315–5_2328.

22. S. S. Yau and F.-C. Chen, An approach to concurrent control flow checking, *IEEE Trans. Softw. Eng.*, no. 2, pp. 126–137, 1980.

23. Z. Alkhalifa, V. S. Nair, N. Krishnamurthy, and J. A. Abraham, Design and evaluation of system-level checks for on-line control flow error detection, *IEEE Trans. Parallel Distrib. Syst.*, vol. 10, no. 6, pp. 627–641, 1999.

24. O. Goloubeva, M. Rebaudengo, M. Sonza Reorda, and M. Violante, Soft-error detection using control flow assertions, in *Proceedings of the 18th IEEE International Symposium on Defect and Fault Tolerance in VLSI Systems*, Boston, MA, November 3–5, 2003, pp. 581–588.

25. O. Goloubeva, M. Rebaudengo, M. Sonza Reorda, and M. Violante, Improved software-based processor control-flow errors detection technique, in *Proceedings of the Annual Reliability and Maintainability Symposium*, Alexandria, VA, 2005, pp. 583–589.

26. J. A. Abraham and K.-H. Huang, Algorithm-based fault tolerance for matrix operations, *IEEE Trans. Comput.*, vol. 100, no. 6, pp. 518–528, 1984.

27. J. A. Abraham and J.-Y. Jou, Fault-tolerant FFT networks, *IEEE Trans. Comput.*, vol. 37, no. 5, pp. 548–561, 1988.

28. B. Randell, System structure for software fault tolerance, *IEEE Trans. Softw. Eng.*, vol.1, no. 2, pp. 220–232, 1975.

29. M. Pignol, DMT and DT2: Two fault-tolerant architectures developed by CNES for COTS-based spacecraft supercomputers, in *12th IEEE International On-Line Test Symposium*, Lake Como, Italy, July 10–12, 2006, pp. 203–212.

30. S. Esposito, C. Albanese, M. Alderighi, F. Casini, L. Giganti, M. L. Esposti, C. Monteleone, and M. Violante, COTS-based high-performance computing for space applications, *IEEE Trans. Nucl. Sci.*, vol. 62, no. 6, 2015, pp. 2687–2694.

2

Soft Errors in Digital Circuits Subjected to Natural Radiation: Characterisation, Modelling and Simulation Issues

Daniela Munteanu and Jean-Luc Autran

CONTENTS

ABSTRACT This chapter surveys soft errors induced by natural radiation on advanced complementary metal oxide semiconductor (CMOS) digital technologies. After introducing the radiation background at ground level (including terrestrial cosmic rays and telluric radiation sources), the chapter describes the main mechanisms of interaction between individual particles (neutrons and charged particles) and circuit materials; it also explains the different steps and production mechanisms of soft errors at device and circuit levels. Then, soft error characterisation using accelerated and real-time tests is surveyed, as well as modelling and numerical simulation issues, with a special emphasis on the Monte Carlo simulation of the soft error rate

(SER). Finally, the radiation response of the most advanced technologies is discussed on the basis of recently published studies focusing on deca-nano-metre bulk, fully depleted silicon-on-insulator (FDSOI) and FinFET families.

2.1 Introduction

As metal oxide semiconductor field-effect transistor (MOSFET) scales have reduced, the integrated circuit (IC) sensitivity to radiation coming from natural space or present in the terrestrial environment has been found to seriously evolve [1–3]. Nowadays, for ultrascaled devices, natural radiation is inducing one of the highest failure rates of all reliability concerns for devices and circuits in the area of nanoelectronics [2]. In particular, ultrascaled memory ICs have been found to be more sensitive to single-event upset (SEU) and digital devices more affected by digital single-event transients (DSETs). This sensitivity to single-event effects (SEEs) is a direct consequence of the reduction of device dimensions and spacing within circuit blocks combined with the reduction of supply voltage and node capacitance, resulting in a decrease of both the critical charge (i.e. the minimum amount of charge required to induce a bit flip) and the sensitive area (i.e. the minimum collection area inside which a given particle can deposit enough charge to induce a bit flip) [2,4].

This chapter examines the issue of soft errors induced by natural radiation at ground level in current and future complementary metal oxide semiconductor (CMOS) digital circuits. The text is structured in four main sections. In Section 2.2, we briefly describe the natural radiation environment at ground and atmospheric levels. The physics and the underlying mechanisms of soft errors at the silicon level are summarised in Section 2.3. We briefly depict the main mechanisms of interaction between single-particle and circuit materials and the electrical response of transistors, cells and complete circuits. Section 2.4 explains soft error characterisation using accelerated and real-time tests, as well as modelling and numerical simulation issues. Finally, Section 2.5 presents and discusses the radiation response of the most advanced technologies, including deca-nanometre bulk, FDSOI and FinFET families.

2.2 Natural Radiation at Ground Level

Soft errors are the result of the interaction of highly energetic particles, such as protons, neutrons, alpha-particles or heavy ions, with the sensitive region(s) of a microelectronic device or circuit. A single event may perturb the device or circuit operation (e.g. reverse or flip the data state of a memory cell, latch, flip-flop, etc.) or definitively damage the circuit (e.g. gate oxide rupture or destructive latch-up events). The problem has been well known for space applications over many years (more than 40 years) and production mechanisms of soft errors in semiconductor devices by energetic protons, electrons and heavy ions well apprehended, characterised and modelled [5]. In a similar way for avionic applications, the interaction of atmospheric high-energy neutrons and protons with electronics has been identified as the major source of soft errors [6]. For the most recent deca-nanometre technologies, the impact of other atmospheric particles produced in nuclear cascade showers on circuits has been clearly demonstrated, in particular low-energy protons [7] and, more recently, low-energy muons [8].

With respect to such high-altitude atmospheric environments, the situation at ground level is slightly different, as illustrated in Figure 2.1: atmospheric particle fluxes are divided by more than two orders of magnitude ($\div$ 300 for neutrons, $\div$ 500 for protons) at sea level with respect to their values at avionics altitude. Such tenuous atmospheric radiation can no longer screen telluric radiation (alpha-particles generated from ultratraces of radioactive contaminants in the CMOS process or packaging materials) that can impact the SER of circuits at ground level [9,10]. As a consequence of these multiple sources of radiation, the accurate modelling and simulation of the SER of circuits at ground level is a rather complex task because one can clearly separate the contribution to SER of atmospheric particles (the external constraint) from the one due to natural alpha-particle emitters present as contaminants in circuit materials (the internal constraint). We briefly detail in the following sections these two natural radiation constraints at ground level.

2.2.1 Atmospheric Radiation

A complex cascade of elementary particles and electromagnetic radiation is generally produced in the earth's atmosphere when a primary cosmic ray (of extraterrestrial origin) interacts with the top atmosphere [11]. The term *cascade* means that the incident particle (generally a proton, nucleus, electron or photon) strikes a molecule in the air so as to produce many high-energy secondary particles (photons, electrons, hadrons and nuclei), which in turn create more particles, and so on.

Among all these secondary particles, neutrons represent the most important part of the natural radiation affecting ground-level susceptibility for current electronics. Because neutrons are not charged, they are very invasive and can penetrate deeply in the circuit materials. They can interact via nuclear reactions with the atoms of the target materials and create (via elastic or inelastic processes) secondary ionising particles. This mechanism is called 'indirect ionisation' and is potentially an important source of errors induced in electronic components. One generally distinguishes thermal neutrons (interacting with ^{10}B isotopes potentially present in circuit materials, but

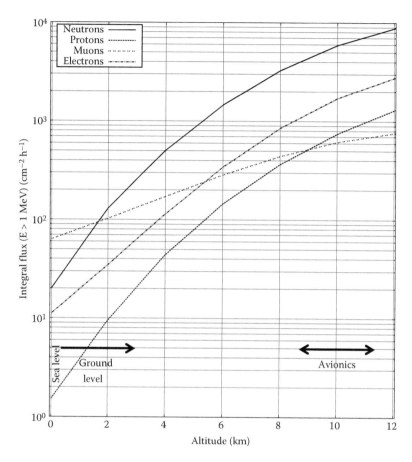

FIGURE 2.1
Integral flux (E > 1 MeV) for atmospheric neutrons, protons, muons and electrons as a function
of the altitude for the geographic coordinates (42°N 72°W) corresponding to New York City.
(Data obtained using the EXPACS model in Sato, T., et al., *Radiat. Res.*, vol. 170, pp. 244–259, 2008.)

progressively removed from technological processes [2]) and high-energy
atmospheric neutrons (up to the gigaelectronvolt scale).

The typical energy distribution of atmospheric high-energy neutrons
(E > 1 MeV) at ground level is shown in Figure 2.2 (lethargic representation).
The integration of this spectrum gives the total neutron flux expressed in
neutrons per square centimetre and per hour. At sea level (New York City),
this flux is equal to ≈20 n/cm^2/h for neutrons above 1 MeV [12]. This value is
reduced to ≈13 n/cm^2/h when integrating the flux above 10 MeV [12].

Atmospheric muons also represent an important part of the natural radiation at ground level [11]. Muons are the products of the decay of secondary
charged pions (p$^\pm$) and kaons (K$^\pm$). In spite of a lifetime of about 2.2 μs, most of
muons survive to sea level due to their ultrarelativistic character. As shown in

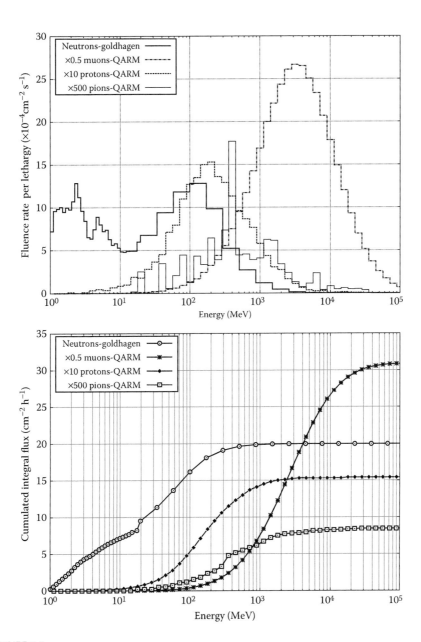

FIGURE 2.2

Fluence rate per lethargy (top) and integral flux (bottom) for high-energy (E > 1 MeV) atmospheric neutron, proton, muon and pion electrons as a function of particle energy. (Experimental numerical data courtesy of P. Goldhagen for neutrons in Gordon, M. S., et al., *IEEE Trans. Nucl. Sci.*, vol. 1, pp. 3427–3434, 2004, and from the QARM numerical model for muons, protons and pions in Quotid Atmospheric Radiation Model (QARM), available at http://82.24.196.225:8080/qarm.)

Figure 2.1, they are the most abundant particles at sea level. The total $(\mu^+ + \mu^-)$ integrated flux above 1 MeV is ≈ 60 $\mu/cm^2/h$ at sea level (Figure 2.2). but only ≈ 600 $\mu/cm^2/h$ at avionics altitude (10 km), as estimated using the EXcel-based Program for calculating Atmospheric Cosmic-ray Spectrum (EXPACS) [13] or quotid atmospheric radiation model (QARM) [14] (Figure 2.2). High-energy physicists are familiar with an order of magnitude of one particle per square centimetre and per minute at sea level for horizontal detectors with an averaged particle energy of 3–4 GeV. But despite this abundance, muons interact very weakly with matter, excepted at low energies (typically below a few gigaelectronvolts) by direct ionisation. The relative importance of low-energy muons in the SER of the most advanced CMOS technologies is discussed in Section 2.5.

In contrast and while strongly interacting with matter, pions are not sufficiently abundant at ground level to induce significant effects in components. Furthermore, for current technologies, the small amount of electrons and gamma-rays susceptible to interact with matter at sea level are not able to disrupt electronics.

Finally, protons, although they interact with silicon as neutrons typically above 50–100 MeV, they are 100 times less numerous than the latter at ground level (see Figure 2.1). Their low abundance at sea level (≈ 1.5 protons/cm^2/h) allows us to consider their impact as negligible compared with that of neutrons. In contrast, for avionics applications, as we said, the number of protons is ≈ 500 times higher than at ground level and they constitute a nonnegligible component of the atmospheric radiation constraint for electronics.

2.2.2 Telluric Radiation Sources

Any terrestrial material contains traces of radioactive atoms, in a wide range varying from a few atoms per thousand for the most active materials to a few atoms per tens of billions for the most purified ones. These natural radioisotopes contained in the earth's crust are the principal natural sources of alpha, beta and gamma radioactivity, but only the alpha-particle emitters present a reliability concern in microelectronics. Beta and gamma processes are indeed not able to deposit a high enough amount of energy to significantly impact the microelectronic circuit operation. On the contrary, alpha-particles (He^{2+}) produced by radioactive decay with typical energies ranging from 1 to 10 MeV can cause a sudden burst of several millions of electrons in silicon over a path of a few tens of microns. This is generally sufficient to induce a transient current that can disturb the operation of a given IC.

Radioactive nuclei can be classified into two categories: the radioactive materials and the radioactive impurities or pollutants [16]. Radioactive materials naturally contain a proportion, generally weak, of alpha-emitter isotope, as, for example, hafnium (^{174}Hf is an alpha-emitter and its natural abundance is 0.162%). The second category corresponds to an unwanted element, that is, unintentionally introduced during the process. This mainly corresponds to uranium and thorium, which have alpha-emitter isotopes in their respective

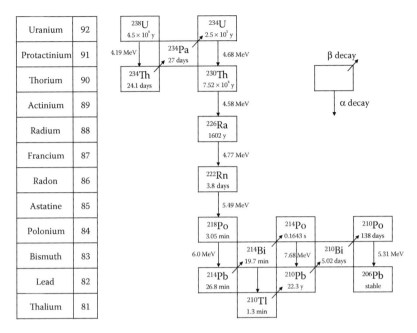

FIGURE 2.3
Uranium 238 radioactive decay chain. The half-life is indicated for each radioisotope of the chain. Energy values of the emitted alpha-particles are also indicated for the different alpha-decays.

disintegration decay chain. ^{232}Th and ^{238}U are widely present in the natural environment and can easily pollute water flow and raw materials used at wafer, packaging and interconnection levels.

Considering the activity of radioisotopes in the calculation of the SER of a circuit thus requires accurate modelling of the alpha-particle source mimicking the presence of these alpha-particle emitters in the circuit materials. Therefore, considering traces of uranium in a given material (e.g. silicon) requires taking into account the complete ^{238}U disintegration shown in Figure 2.3. This chain is composed of 14 daughter nuclei with 8 alpha-particle emitters [17]. The energies of these alpha-particles range from 4.20 to 7.68 MeV; their corresponding ranges in silicon respectively vary from 19 to 46 μm, which is much larger than the characteristic dimensions of a memory bitcell in current circuits, for example.

2.3 Soft Error Production Mechanisms at the Silicon Level

The physical underlying mechanisms related to the production of soft errors in microelectronic devices schematically consist of successive steps,

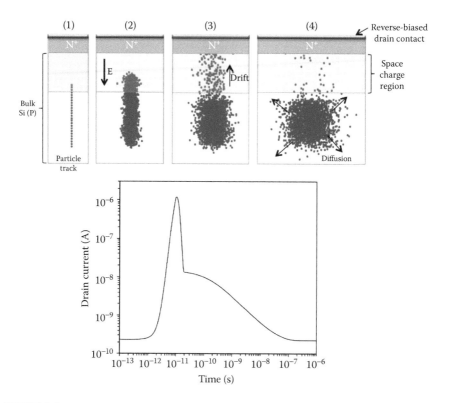

FIGURE 2.4
Charge generation, transport and collections phases in a reverse-biased junction and the resultant current pulse caused by the passage of a high-energy ion. (After Glorieux, M., et al. *IEEE Trans. Nucl. Sci.*, vol. 61, no. 6, pp. 3527–3532, 2014.)

illustrated in Figure 2.4 in the case of an alpha-particle striking a reverse-biased n+/p junction [18]: (1) the charge deposition by the energetic particle within the sensitive region, (2) the transport of the released charge into the device and (3 and 4) the charge collection in the active region of the device. In the following, we succinctly describe these different mechanisms at the origin of SEUs or SETs in digital circuits.

2.3.1 Charge Deposition (or Generation)

When an energetic charged particle strikes the device, an electrical charge along the particle track can be deposited by one of the following mechanisms: direct ionisation by the interaction with the material or indirect ionisation, by secondary particles issued from nuclear reactions with the atoms of the struck material. Direct ionisation by heavy ions ($Z \geq 2$) of the space environment is particularly important. These interact with the target material mainly by inelastic interactions and transmit a large amount of energy to the electrons of the struck atoms. These electrons produce a cascade of secondary electrons, which thermalise and create electron–hole pairs along the particle path (Figure 2.4 [1]). In a semiconductor or insulator, a large amount of the deposited energy is thus converted into electron–hole pairs, the remaining energy being converted into heat and a very small quantity into atomic displacements. It was shown experimentally that the energy necessary for the creation of an electron–hole pair depends on the material bandgap. In a microelectronics silicon substrate, one electron–hole pair is produced for every 3.6 eV of energy lost by the ion. Other particles, such as the neutrons of the terrestrial environment, do not interact directly with the atomic electronics of the target material and so do not ionise the matter on their passage. However, these particles should not be neglected, because they can produce SEE due to their probability of nuclear reaction with the atoms of materials that compose the microelectronic devices. This mechanism is called indirect ionisation. The charged products resulting from a nuclear reaction can deposit energy along their tracks, in the same manner as that of direct ionisation. Since the creation of the column of electron–hole pairs of these secondary particles is similar to that of ions, the same general models and concepts can be used.

2.3.2 Charge Transport

When a charge column is created in the semiconductor by an ionising particle, the released carriers are quickly transported and collected by elementary structures (e.g. p-n junctions). The transport of charge relies on two main mechanisms (Figure 2.4 [2] to [4]): the charge drift in regions with an electric field and the charge diffusion in neutral zones. The deposited charges can also recombine with other mobile carriers existing in the lattice.

2.3.3 Charge Collection

The charges transported in the device induce a parasitic current transient (Figure 2.4, bottom), which can induce disturbances in the device and associated circuits. The devices most sensitive to ionising particle strikes are generally devices containing reverse-biased p-n junctions, because the strong electric field existing in the depletion region of the p-n junction allows a very efficient collection of the deposited charge. The effects of ionising radiation are different according to the intensity of the current transient, as well as the number of circuit nodes impacted. If the current is sufficiently important, it can induce permanent damage on gate insulators (single-event gate rupture [SEGR]) or the latch-up (single-event latch-up [SEL]) of the device. In usual low-power circuits, the transient current may generally induce only an eventual change of the logical state (cell upset).

2.4 Soft Error Characterisation and Modelling Issues

2.4.1 Soft Error Characterisation Using Accelerated and Real-Time Tests

To predict the impact of natural radiation on the behaviour of electronics and to (statistically) estimate (i.e. measure) its radiation-induced SER, three main experimental approaches can be envisaged [9,19], excluding modelling and simulation methods that can be used, under certain conditions (i.e. when correctly calibrated), as predictive tools (see Section 2.4.2).

The first one, called 'field testing', consists in collecting errors from a large number of finished products already on the market. The SER value is evaluated *a posteriori* from the errors experienced by the consumers themselves; it generally takes several years after the introduction of the product on the market. This method is not adapted to upstream reliability studies performed during the cycle of product development and will not be considered in the following.

The second method, called accelerated soft error rate (ASER), consists of using intense particle beams or sources chosen for their capability to mimic the atmospheric (neutron) spectrum or to generate alpha-particles within the same energy range as the alphas emitted by radioactive contaminants [9,19]. This ASER method is fast (data can be obtained in a few hours or days, instead of months or years for the other methods), *a priori* easy to implement and only requires a few functional chips to estimate the SER. This allows the manufacturer to perform such radiation tests relatively early in the production cycle. Another major and growing advantage is its capability to quantify from very large statistics (cumulated number of events) the importance of multiple-cell (MCU) and multiple-bit upsets in the radiation response of ICs fabricated in technological nodes, typically below 65 nm. But data can be potentially tainted by experimental artefacts (more or less well controlled according to the facility, the experimental setup or other various experimental conditions). As a direct consequence, ASER results must be extrapolated to use conditions and several different radiation sources must be used to ensure that the estimation accounts for soft errors induced by both alpha-particle and cosmic-ray neutron events.

The third method consists of exposing a given device (or a large number of identical devices) to terrestrial radiation over a sufficiently long period (weeks or months) in order to achieve adequate statistics on the number of accumulated errors and then on the SER value. This method is called real-time soft error rate (RTSER) test [20–23] or unaccelerated testing. In this method, as in accelerated testing, the intensity of the natural radiation can be increased by deploying the test in altitude (at least for neutrons). However, the acceleration factor (AF) (i.e. the ratio of the neutron integrated flux at the test location divided by its reference value in New York City [19]) has nothing to do with those reached in accelerated tests. Considering an equivalence of the radiation background composition in altitude and at sea level [19], typical AF values between 5 and 20, as a function of the test location on earth, can be expected. Devices have thus to be tested for a long enough period of time (months or years) until enough soft errors have been accumulated to give a reasonably confident estimate of the SER. The main advantage of RTSER tests is that they provide a direct measurement of the 'true' SER that does not require intense radiation sources and extrapolations to use conditions. The major drawbacks of this method concern the cost of the system (which has to be capable of monitoring a very large number of devices at the same time) and the long duration of the experiment.

Figure 2.5 illustrates this RTSER method in the case of a mountain altitude (AF ≈ 6) experiment conducted in 2011–2014 on the Altitude SEE Test European Platform (ASTEP). In this experiment [23], 7168 Mbits of CMOS 40 nm static random access memory (SRAM) have been continuously

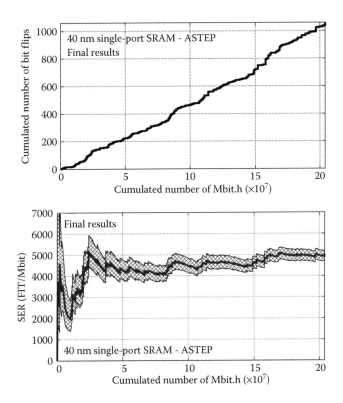

FIGURE 2.5
RTSER characterisation of 40 nm single-port SRAMs exposed to natural radiation on the ASTEP platform. Top: Cumulative number of bit flips vs. test duration. Test has been conducted under nominal conditions: $V_{DD} = 1.1$ V, room temperature, standard checkerboard test pattern. Bottom: Bit flip SER (FIT/MBit) vs. test duration calculated from the above data. The 90% confidence interval is also indicated (hatched area).

monitored using a dedicated automatic test equipment and under nominal conditions (room temperature, core voltage $V_{DD} = 1.1$ V). Figure 2.5 (top) shows the cumulative distribution of bit flips during this RTSER experiment. A total of 1021 bit flips have been detected in more than 27,000 h of measurements. Single-bit upsets (SBUs) only represents 217 bit flips, the 804 remaining being due to 158 MCU events. The average multiplicity (i.e. number of bit flips involved in the same event, also called MCU size) of MCU events is then equal to $804/158 \approx 5.1$ for this technology (compared with 2.0 for 130 nm and 3.1 for 65 nm in the same conditions, i.e. on ASTEP), with rare events up to 15 (3 events), 16 (3 events), 21 (1 event) and 22 (1 event) of multiplicity detected during these 3 years of experiment. As a direct consequence, Figure 2.5 (top) also shows that these MCUs with large multiplicities introduce irregular staircases on the bit flip distribution, then justifying *a posteriori* such a very long experiment duration (more than 3 years) for correctly extracting the averaged SER value. Figure 2.5 (bottom) precisely shows the convergence of such a SER value as a function of time. The total bit flip SER resulting from this test is reported in Table 2.1, with other values corresponding to neutron and alpha-particle accelerated tests. The extracted SER value at sea level for the same technology (during a complementary test performed in Marseille with the same test setup) is found to be in nice agreement with the total SER value expressed for New York City conditions and deduced from RT experiments in altitude corrected from the alpha-SER value (separately estimated using an accelerating test with a ^{241}Am solid source).

TABLE 2.1

Summary of the RTSER Experiment (Figure 2.5) Conducted on the ASTEP Platform during the Period 2011–2014 and Involving More Than 7 Gbits of CMOS Bulk 40 nm SRAM under Test

Values Deduced From	SER Component	Bit Flip SER (FIT/Mbit) at New York City 40 nm SRAM
Real-time experiment in altitude (ASTEP platform)	Neutron	835
	Alpha	471 (taking the value deduced from accelerated test)
	Measured (neutron + alpha)	1306 $\updownarrow$
Real-time experiment at sea level (Marseille)	Neutron + alpha	1348
Accelerated tests using intense sources	Neutron (facility)	880 (TRIUMF)
	Alpha (^{241}Am)	471 (extrapolated for an emissivity of 0.001 alpha/cm²/h)
	Total (neutron + alpha)	1351

Source: Autran, J. L., et al., Altitude and underground real-time SER testing of SRAMs manufactured in CMOS bulk 130, 65 and 40 nm, presented at *2014 IEEE Radiation Effects Data Workshop (REDW)*, Paris, July 14–18, 2014.

2.4.2 Modelling Issues

Modelling and simulating the effects of ionising radiation has long been used for better understanding radiation effects on the operation of devices and circuits. In the last two decades, due to substantial progress in simulation codes and computer performances that reduce computation times, simulation has acquired significantly increased interest [24,25]. Due to its predictive capability, simulation offers the possibility to reduce radiation experiments and to test hypothetical devices or conditions, which are not feasible (or not easily measurable) by experiments. The continuous reduction of the feature size in microelectronics requires increasingly complicated and time-consuming manufacturing processes. Therefore, a systematic experimental investigation of the radiation effects of new ultrascaled devices or emerging devices with alternative architecture (such as multiple-gate or silicon nanowire transistors) is difficult and expensive. Since computers are today considerably cheaper resources, simulation is becoming an indispensable tool for the device engineer, not only for device optimisation, but also for specific studies, such as the device sensitivity when subjected to ionising radiation. Last but not least, the understanding of the soft error mechanisms in ultrascaled devices and the prediction of their occurrence under a given radiation environment are of fundamental importance for certain applications requiring a very high level of reliability and dependability [9].

2.4.2.1 Device-Level Modelling Approach

Simulation of radiation effects at the device level aims to describe both the device (physical construction and electrical behaviour) and its operation in a given radiation environment or when it is subjected to a particular type of radiation. Two methods can be used for this purpose: the device numerical simulation (technology computer-aided design [TCAD]) and the use of compact models (which are later included in circuit-level SPICE-like simulations). TCAD numerical modelling concerns both the simulation of the manufacturing process and the simulation of the device electrical operation. Concerning this last point, the electrical simulator solves the main differential physical equations, such as the Poisson equation and the transport and continuity equations. The main inconvenience of this type of simulation is the computation time, which can be very important depending on the simulation domain size and the equations considered for the transport (drift diffusion or hydrodynamic). But the significant advantage of TCAD is the ability to access internal quantities of the simulation (which cannot be measured), which substantially facilitates the fine understanding of the physical and electrical mechanisms taking place in the device. In contrast to TCAD, compact models are based on analytical formulae that describe the static or dynamic electrical behaviour of the elementary devices constituting the circuit. They literally constitute a bridge between the device (which is itself closely related to technology) and the circuit design.

2.4.2.2 Circuit-Level Modelling Approaches

Three main modelling approaches are used for the simulation of SEEs at the circuit level: circuit-level simulation and mixed-mode and full numerical simulation in the three-dimensional (3D) device domain. The latter currently remains the most accurate solution for studying SEE in circuits since it numerically models and solves the entire impacted subcircuit in the 3D device domain. This was possible only recently (typically in the past decade), due to the enhancement of computer performances (central processing unit [CPU] clock speed and memory resources), which reduced the computational time.

2.4.2.3 Monte Carlo Simulation Tools

Full Monte Carlo–based physical simulations of the SER provide a very powerful way to bring much more detailed physics to bear on the process of error rate prediction than has heretofore been possible with models and analytical computations [26–28]. A limited number of code developments have been reported in the literature in the domain of SEEs [27,28], in so far as the complete simulation chain is complex (due to its multiscale and multiphysics character), and require long-term developments. Schematically, Monte Carlo simulation codes solve the radiation problem in two main steps, the interaction of radiation with the device and the subsequent motion of charges, and resulting changes in nodal currents and voltages, within the device or circuit. To illustrate this approach, Figure 2.6 shows the schematics of the Tool Suite for Radiation Reliability Assessment (TIARA-G4) code, a proprietary Monte Carlo simulation platform conjointly developed by STMicroelectronics and Aix-Marseille University (IM2NP laboratory) [28]. The code is a general-purpose Monte Carlo simulation program written in C++ and based on the Geant4 toolkit [29] for modelling the interaction of several particles (including neutrons, protons, muons, alpha-particles and heavy ions) with various architectures of electronic circuits. The first step of the simulation flow is to construct a model of the simulated circuit from Geant4 geometry classes and libraries of elements and materials. In the framework of Geant4, the circuit under simulation is considered the 'particle detector'. The structure creation in TIARA is based on 3D circuit geometry information extracted from Graphic Database System (GDS) formatted data classically used in the integrated circuit computer-aided design (IC-CAD) flow of semiconductor circuit manufacturing. The real 3D geometry is simplified since it is essentially based on the juxtaposition of boxes of different dimensions, each box being associated with a given material (silicon, insulator, metal, etc.) or doped semiconductor (p-type, n-type).

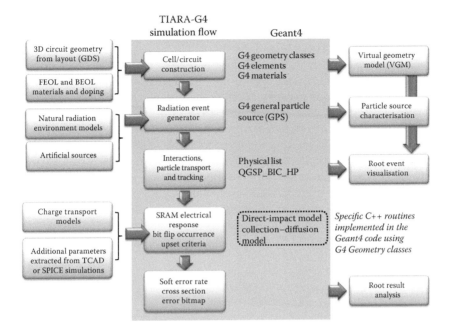

FIGURE 2.6
Schematics of the TIARA-G4 simulation flow showing the different code inputs and outputs and the links with Geant4 classes, libraries, models or external modules and visualisation tools.

To numerically generate the particles with the spectral, spatial and angular distributions mimicking all the characteristics of the natural background, TIARA-G4 uses the Geant4 General Particle Source (GPS) [30], which is part of the Geant4 distribution. The module allows the user to define all the source parameters, in particular the energy of the emitted particles from a given energy distribution defined in a separate input file. Once an incident particle has been numerically generated with the radiation event generator, the Geant4 simulation flow computes the interactions of this particle with the target (the simulated circuit) and transports step by step the particle and all the secondary particles eventually produced inside the world volume (the largest volume containing, with some margins, all other volumes contained in the circuit geometry). The transport of each particle occurs until the particle loses its kinetic energy to zero, disappears by an interaction or comes to the end of the world volume. During the simulation flow, TIARA-G4 automatically generates output files describing all the particle interaction events and allowing the tracking of all the secondary particles impacting different sensitive volumes of the circuit; these files can be used later for event visualisation or postprocessing. Finally, TIARA-G4 automatically generates ROOT [31] scripts for the visualisation of interaction events. These separate scripts directly import geometry and event data from a collection of files saved on the machine hard disk during simulation. At the end of a simulation sequence, TIARA-G4 examines the tracks of all the charged particles involved in this step (including eventually the track of the incident primary particle if it is charged) and determines the complete list of the different silicon volumes (drains, Pwells, Nwells, substrates, etc.) traversed by these particles. Taking into account the type and architecture of the circuit under simulation (SRAM, flip-flop, flash memory, etc.), TIARA-G4 computes its electrical response induced by the incident particle strike. At this level, it is possible to develop a collection of electrical models and call a specific module from a dedicated user's library. Finally, at the end of the simulation flow, the last module of the TIARA-G4 code evaluates the SER of the considered circuit from the total number of errors obtained during the simulation run, the circuit dimensions and the characteristics of the simulated source of particles. It also generates additional files for the cross section distribution and error bitmap.

2.5 Radiation Response of Advanced Technologies

While CMOS technologies continue to shrink in a 'more Moore' perspective, new risks are arising with scaling for SEEs. Indeed, the SEE susceptibility of advanced technologies is expected to evolve under the influence of several factors since extrinsic radiation does not scale down. The most important are [32,33]: (1) the reduction of the critical charge, (2) the reduction of the per-bit cross sections presented to ionising particles, (3) the reduction of the energy deposition volumes traversed by the particle at the front-end level, (4) the increase of the particle region of influence at the layout level and (5) the amplification of parasitic effects as a function of device architecture considered.

The critical charge (Q_{crit}) corresponds to the minimum charge disturbance needed to flip a logic level. Figure 2.7 shows the drastic reduction of this charge when pushing device integration for bulk and SOI technologies. Current technologies (28 and 22 nm nodes) are operating in a Q_{crit} regime of a few thousand electrons, or hundreds of atto-coulomb (aC), a value well below the amount of charge deposited by a singly charged ionising particle in silicon. Consequently, logic circuits with low node capacitances and operating under low power supply voltages will be more susceptible to suffer from soft errors than previous generations of circuits. Reduced Q_{crit} also leads to expand the spectrum of particles to which a circuit is sensitive. Although a very low ionising power, atmospheric muons and low-energy protons are now capable of creating upsets in the most integrated technologies. Low-energy

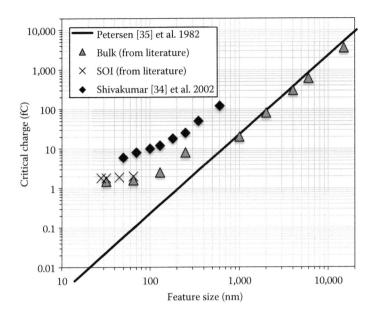

FIGURE 2.7
Critical charge scaling as a function of feature size. (Data from Massengill, L. W., et al., in *2012 IEEE International Reliability Physics Symposium (IRPS)*, Anaheim, CA, April 15–19, 2012, pp. 3C.1.1–3C.1.7; Shivakumar, P., et al., in *Proceedings of the International Conference on Dependable Systems and Networks*, Bethesda, MD, June 23–26, 2002, pp. 389–398; Petersen, E. L., et al., *IEEE Trans. Nucl. Sci.*, vol. 29, no.6, pp. 2055–2063, 1982.)

protons (<10 MeV) are primarily a concern for space applications, although they can also be generated by sea-level neutrons colliding with silicon nuclei or by the local environment at ground level (see Section 2.1). Recent proton testings have demonstrated their importance to SER for bulk SRAMs in 40 and 28 nm. In a similar way, atmospheric muons, which are the most preponderant charged particles at sea level (see Figure 2.1), are also susceptible to represent a new radiation threat for deca-nanometre technologies, specially operated at ultralow voltages. The uncertainty of the terrestrial muon flux below a few hundred megaelectronvolts does not yet enable an accurate estimation of their direct SER impact at ground level, even if 1000× projections have been made [29]. Moreover, when negative muons have lost their kinetic energy and stop, about 35% spontaneously decay. The remaining 65% are captured, producing in silicon unstable aluminium that can de-excite by evaporation. This additional mechanism should be a potential source of soft errors for future technologies [3].

Other important issues have been identified for bulk or SOI deca-nanometre technologies [32]: the growing importance of telluric rays (alpha-particles generated from traces of radioactive contaminants in IC materials), the lesser efficiency of classical mitigation techniques, the enhanced total ionising dose (TID) due to X- or gamma-rays (also protons and electrons), and so forth. We try to summarise in the following the most important key points of these issues for current commercial bulk and SOI technologies.

- *Bulk technologies*: The evolution of the measured single-bit SRAM SER as a function of technology nodes shows that the SER per bit has peaked at the 130 nm technology node and since then has been decreasing. This current reduction tendency can be explained by the dramatic reduction of the charge-collection volumes due to the geometrical scaling compensating lower critical charges. At chip level, since the cells are more densely packed with scaling, the charge from a single particle is more easily shared among several cells. This reduces the amount of charge effectively collected by an individual cell. The net effect is an improved per-bit SER of scaled technologies, as observed in dynamic random access memories (DRAMs), dynamic logic and SRAMs, but not for latches and combinatorial logic [32]. Finally, at circuit or system-on-chip levels, SER keeps increasing with the growing amount of memories and latches and with a particular stronger SER impact on latches. The decrease of supply voltages for higher power efficiency (e.g. for cloud computing) is an aggravating factor and will constitute a grand reliability challenge.

- *SOI technologies*: The sensitive volume traversed by an ionising particle is even further reduced with the isolation barriers, resulting in a stronger intrinsic SER resilience in SOI compared with bulk. This total isolation of device also removes the parasitic thyristor at the origin of SELs. SOI technologies are then immune to SEL by construction, and this immunity is also verified for hybrid bulk devices in FDSOI 28 nm [32]. However, a stronger parasitic bipolar in SOI can degrade SER despite small critical charges and small sensitive volumes. Connecting the partially depleted (PD)-SOI internal body to either source or ground greatly improves SER, but induces area penalty. In FDSOI 28 nm, the bipolar gain has been measured very low (<3 in the worst case), and full benefit can be taken from the ultrasmall sensitive volume, thus strongly minimising SER. Recent results have experimentally demonstrated a 110× reduction factor in the high-energy neutron SER in FDSOI 28 nm compared with bulk 28 nm, with the lowest failure-in-time (FIT) rate ever observed for SRAMs (<10 FIT/Mbit) [32].

To conclude, Table 2.2 summarises the expected SER, SEL and TID performances between CMOS bulk, FinFET and SOI technologies. This table has been compiled from references cited in [32] and offers a global overview of the performances, drawbacks and main challenges for the upcoming technologies. Certain issues for ultrascaled SOI devices or FinFET architectures should be investigated in depth in future studies.

TABLE 2.2

Expected SER, SEL and TID Performances between CMOS Bulk, FinFET and SOI Technologies

With Respect to Bulk CMOS	PD-SOI	Bulk FinFET	SOI-FinFET Experimental	UTBB-FD-SOI
Critical charge, minimal charge to upset	0.1 fC	<0.1 fC	<0.1 fC	<0.1 fC
Sensitive volume, charge deposition and collection	Small	Very small	Very small	Ultrasmall
Parasitic bipolar, charge amplification	Significant without body ties, <20	Ultralow substrate tied to body	Low, ~2–8	Ultralow, <3
Alpha/ neutron-SER	÷5 to ÷20	÷2.5 to ÷3.5	÷10	÷100
Muon-SER	New SER risk (1000×)			
Thermal neutron-SER	New SER risk (2×)			
Low-energy proton-SER	New SER risk (to be evaluated)			
SEL, ion-induced latch-up	Immune by construct	No data yet in literature	Immune by construct	Immune, including hybrid devices
TID, gamma- and X-rays	Mega-rad with body ties/taps	100s of kilo-rad with large fins	Mega-rad with narrow fins	100s of kilo-rad

Source: Adapted from Roche, P., et al. in Technical Digests of the *IEEE International Electron Device Meeting (IEDM)*, Washington, DC, December 9–11, 2013, pp. 766–769.

Note: UTBB, ultrathin body and box.

2.6 Conclusion

Radiation effects on CMOS technologies at ground and atmospheric levels have deeply evolved in the last decade, typically since the introduction of the bulk 130 nm technology node. The geometrical scaling combined with the core voltage reduction and the operation frequency increase has induced the emergence of new mechanisms (charge sharing, bipolar amplification and carrier channelling in wells) and enlarged the region of influences of particles in circuits (multinode charge collection). All these effects have profoundly modified the radiation response of devices and circuits. These latter have become more sensitive to tenuous radiation (alpha-particle emitters, low-energy protons and atmospheric muons). Understanding evolving risks of SEEs for current and future CMOS technologies requires more than ever an in-depth knowledge of complex domains, including the natural radiation environment, the particle–matter interactions (extended to new particles as muons) and the physics of soft error mechanisms in new device architectures and circuits (extended planar bulk, FDSOI and FinFET). The long-term objective of all the works performed in this domain should be to progress towards a predictive simulation of SEEs in future nanoelectronic circuits.

References

1. P. E. Dodd and L. W. Massengill, Basic mechanisms and modeling of single-event upset in digital microelectronics, *IEEE Trans. Nucl. Sci.*, vol. 50, no. 3, pp. 583–602, 2003.
2. R. C. Baumann, Radiation-induced soft errors in advanced semiconductor technologies, *IEEE Trans. Device Mater. Reliab.*, vol. 5, no. 3, pp. 305–316, 2005.
3. J. L. Autran and D. Munteanu, *Soft Errors: From Particles to Circuits,* Taylor & Francis Group/CRC Press, Boca Raton, FL, 2015.
4. S. Mitra, P. Sanda, and N. Seifert, Soft errors: Technology trends, system effects and protection techniques, presented at IEEE International On-Line Testing Symposium (IOLTS), Crete, Greece, July 8–11, 2007.
5. R. D. Schrimpf and D. M. Fleetwood (Eds.), *Radiation Effects and Soft Errors in Integrated Circuits and Electronic Devices*, World Scientific Publishing, Singapore, 2004.
6. E. Normand, Single event upset at ground level, *IEEE Trans. Nucl. Sci.*, vol. NS-43, no. 6, pp. 2742–2750, 1996.
7. E. H. Cannon, M. Cabanas-Holmen, J. Wert, T. Amort, R. Brees, J. Koehn, B. Meaker, and E. Normand, Heavy ion, high-energy, and low-energy proton SEE sensitivity of 90-nm RHBD SRAMs, *IEEE Trans. Nucl. Sci.*, vol. 56, no. 7, pp. 3493–3499, 2010.
8. B. D. Sierawski, M. H. Mendenhall, R. A. Reed, M. A. Clemens, R. A. Weller, R. D. Schrimpf, E. W. Blackmore, et al., Muon-induced single event upsets in deep-submicron technology, *IEEE Trans. Nucl. Sci.*, vol. 57, no. 6, pp. 3273–3278, 2010.
9. J. F. Ziegler and H. Puchner, *SER – History, Trends and Challenges,* Cypress Semiconductor, San Jose, CA, 2004.
10. J. L. Autran, D. Munteanu, P. Roche, G. Gasiot, S. Martinie, S. Uznanski, S. Sauze, et al., Soft-errors induced by terrestrial neutrons and natural alpha-particle emitters in advanced memory circuits at ground level, *Microelectron. Reliab.*, vol. 50, pp. 1822–1831, 2010.
11. L. I. Dorman, *Cosmic Rays in the Earth's Atmosphere and Underground*, Kluwer Academic, Dordrecht, the Netherlands, 2004.
12. J. L. Autran, S. Serre, S. Semikh, D. Munteanu, G. Gasiot, and P. Roche, Soft-error rate induced by thermal and low energy neutrons in 40 nm SRAMs, *IEEE Trans. Nucl. Sci.*, vol. 59, no. 6, pp. 2658–2665, 2012.
13. T. Sato, H. Yasuda, K. Niita, A. Endo, and L. Sihver, Development of PARMA: PHITS based analytical radiation model in the atmosphere, *Radiat. Res.*, vol. 170, pp. 244–259, 2008.

14. Quotid Atmospheric Radiation Model (QARM), available at http://82.24.196.225:8080/qarm.
15. M. S. Gordon, P. Goldhagen, K. P. Rodbell, T. H. Zabel, H. H. K. Tang, J. M. Clem, and P. Bailey, Measurement of the flux and energy spectrum of cosmic-ray induced neutrons on the ground, *IEEE Trans. Nucl. Sci.*, vol. 1, pp. 3427–3434, 2004.
16. F. Wrobel, J. Gasiot, and F. Saigné, Hafnium and uranium contributions to soft error rate at ground level, *IEEE Trans. Nucl. Sci.*, vol. 55, pp. 3141–3145, 2008.
17. S. Martinie, J. L. Autran, D. Munteanu, F. Wrobel, M. Gedion, and F. Saigné, Analytical modeling of alpha-particle emission rate at wafer-level, *IEEE Trans. Nucl. Sci.*, vol. 58, no. 6, pp. 2798–2803, 2011.
18. M. Glorieux, J. L. Autran, D. Munteanu, S. Clerc, G. Gasiot, and P. Roche, Random-walk drift-diffusion charge-collection model for reverse-biased junctions embedded in circuits, *IEEE Trans. Nucl. Sci.*, vol. 61, pp. 3527–3532, 2014.
19. JEDEC, Measurement and reporting of alpha particles and terrestrial cosmic ray-induced soft errors in semiconductor devices, JESD89, JEDEC Solid State Technology Association, Arlington, VA, 2006.
20. J. L. Autran, P. Roche, S. Sauze, G. Gasiot, D. Munteanu, P. Loaiza, M. Zampaolo, and J. Borel, Altitude and underground real-time SER characterization of CMOS 65nm SRAM, *IEEE Trans. Nucl. Sci.*, vol. 56, no. 4, pp. 2258–2266, 2009.
21. J. L. Autran, S. Serre, D. Munteanu, S. Martinie, S. Semikh, S. Sauze, S. Uznanski, G. Gasiot, and P. Roche, Real-time soft-error testing of 40nm SRAMs, in *2012 IEEE International Reliability Physics Symposium*, Anaheim, CA, April 15–19, 2012, pp. 3C.5.1–3C.5.8.
22. J. L. Autran, D. Munteanu, G. Gasiot, and P. Roche, Real-time soft-error rate measurements: A review, *Microelectron. Reliab.*, vol. 54, pp. 1455–1476, 2014.
23. J. L. Autran, D. Munteanu, S. Sauze, G. Gasiot, and P. Roche, Altitude and underground real-time SER testing of SRAMs manufactured in CMOS bulk 130, 65 and 40 nm, presented at *2014 IEEE Radiation Effects Data Workshop (REDW)*, Paris, July 14–18, 2014.
24. P. E. Dodd, Physics-based simulation of single-event effects, *IEEE Trans. Device Mater. Reliab.*, vol. 5, no. 3, pp. 343–357, 2005.
25. D. Munteanu and J. L. Autran, Modeling and simulation of single-event effects in digital devices and ICs, *IEEE Trans. Nucl. Sci.*, vol. 55, no. 4, pp. 1854–1878, 2008.
26. R. A. Weller, R. D. Schrimpf, R. A. Reed, M. H. Mendenhall, K. M. Warren, B. D. Sierawski, and L. W. Massengill, Monte Carlo simulation of single event effects, presented at Radiation and Its Effects on Components and Systems (RADECS), Brugge, Belgium, September 14–18, 2009.
27. R. A. Reed, R. A. Weller, A. Akkerman, J. Barak, S. Duzellier, C. Foster, M. Gaillardin, et al., Anthology of the development of radiation transport tools as applied to single event effects, *IEEE Trans. Nucl. Sci.*, vol. 60, no. 3, pp. 1876–1911, 2013.
28. P. Roche, G. Gasiot, J. L. Autran, D. Munteanu, R. A. Reed, and R. A. Weller, Application of the TIARA radiation transport tool to single event effects simulation, *IEEE Trans. Nucl. Sci.*, vol. 61, no. 3, pp. 1498–1500, 2014.
29. S. Agostinelli, J. Allison, K. Amako, and D. Zschiesche, Geant4 – A simulation toolkit, *Nucl. Instrum. Methods Phys. Res. A*, vol. 506, pp. 250–303, 2003.

30. Geant4 General Particle Source, available at https://geant4.web.cern.ch/geant4/UserDocumentation/UsersGuides/ForApplicationDeveloper/html/ch02s07.html.
31. ROOT, An object oriented framework for large scale data analysis, available at http://root.cern.ch.
32. P. Roche, J. L. Autran, G. Gasiot, and D. Munteanu, Technology downscaling worsening radiation effects in bulk: SOI to the rescue, in *IEEE International Electron Device Meeting (IEDM)*, Washington, DC, December 9–11, 2013, pp. 766–769.
33. L. W. Massengill, B. L. Bhuva, W. T. Holman, M. L. Alles, and T. D. Loveless, Technology scaling and soft error reliability, in *2012 IEEE International Reliability Physics Symposium (IRPS)*, Anaheim, CA, April 15–19, 2012, pp. 3C.1.1–3C.1.7.
34. P. Shivakumar, M. Kistler, S. W. Keckler, D. Burger, and L. Alvisi, Modeling the effect of technology trends on the soft error rate of combinational logic, in *Proceedings of the International Conference on Dependable Systems and Networks*, Bethesda, MD, June 23–26, 2002, pp. 389–398.
35. E. L. Petersen, P. Shapiro, J. H. Adams Jr., and E. A. Burke, Calculation of cosmic-ray induced soft upsets and scaling in VLSI devices, *IEEE Trans. Nucl. Sci.*, vol. 29, no. 6, pp. 2055–2063, 1982.

3

Simulation of Single-Event Effects on Fully Depleted Silicon-on-Insulator (FDSOI) CMOS

Walter Calienes Bartra, Andreas Vladimirescu and Ricardo Reis

CONTENTS

ABSTRACT   This chapter is dedicated to single-event effects on fully-depleted silicon-on-insulator (FDSOI) complementary metal oxide semiconductor (CMOS). Circuits using two technological nodes are simulated against heavy-ion impact effects: 32 nm bulk and 28 nm FDSOI. The simulations were done using technology computer-aided design (TCAD) tools.

These devices were used to design six-transistor static random access memory (6T SRAM) cells. These memories were simulated to observe single-event upset (SEU) faults due to heavy-ion impacts in different angles and locations. The 6T SRAM cells designed with FDSOI technology were more resilient against the heavy-ion impacts than the bulk ones.

3.1 Introduction

The continuous scaling of transistors is allowing an increase in the number of components on a chip and also a reduction of voltage values defining a 'logical' one. This voltage reduction also makes the circuits more sensitive to radiation effects, as the needed charge to cause a bit to flip is reduced. This chapter presents simulations of single-event effects in fully depleted silicon-on-insulator (FDSOI) transistors and static random access memory (SRAM) cells, comparing these effects with the ones in a traditional bulk complementary metal oxide semiconductor (CMOS) technology. A comparison of resilience with heavy-ion impacts on the drain region between a 32 nm bulk CMOS transistor, a 28 nm FDSOI transistor and a 28 nm high-K FDSOI transistor is presented. The impacts were performed in different transistor locations at different impact angles, whereas previous works considered the impact just at a 0° angle. This comparison was performed with the device in the off state using two-dimensional (2D) technology computer-aided design (TCAD) simulations.

3.2 Fundamentals of the Single-Event Effects

Space and ground environments are reached by a lot of particles created by solar, cosmic or terrestrial activities. These particles can be charged particles (such as electrons, protons or heavy ions) or electromagnetic radiation (such as X-ray and gamma photons). When one of these particles funnels through the silicon die, it loses energy due to an electron–hole pair production. Protons and neutrons can be produced by nuclear reactions, and they can ionise silicon in a similar manner. The particle ionises the silicon in its track, as shown in Figure 3.1. In summary, the basic transient mechanism due to a particle impact can be described in three steps: (1) charge deposition, (2) charge transport and (3) charge collection (Munteanu and Autran, 2008). These phenomena are due to the photocurrents generated in silicon when it is hit by particles or radiation (Calienes and Reis, 2011).

The charges created by particle impact vary with the type of the particle, the hit angle θ and the impact location (Messenger, 1982). These charges

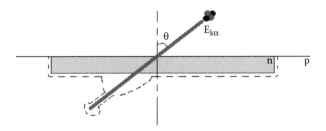

FIGURE 3.1
Alpha-particle with kinetic energy $E_k\alpha$ impacts an inverse-biased n-p silicon junction with angle θ. The dotted line represents a metallurgic junction electric field.

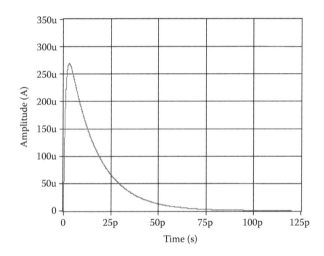

FIGURE 3.2
Transient current simulation. The total CC in this case is 31.5 fC. (From Calienes, W., and Reis, R., SET and SEU simulation toolkit for LabVIEW, presented at Proceedings of European Conference on Radiation and Its Effects on Components and Systems [RADECS], Seville, Spain, September 19, 2011.)

produce an additional transient current $I_p(t)$ and an abnormal charge Q_p in the silicon structure. The model of this transient is summarised in the follow equations:

$$I_p(t) = I_0(exp(-t/\tau_F) - exp(-t/\tau_R)) \tag{3.1}$$

$$Q_p = I_0(\tau_F - \tau_R) \tag{3.2}$$

where I_0 is the maximum current due the generated charges, τ_R is the collection time constant of the junction and τ_F is the time constant to establish the ion track. Figure 3.2 presents an example of transient current simulation using Equation 3.1, with $I_0 = 350$ µA, $\tau_R = 10$ ps and $\tau_F = 100$ ps. The transient current $I_p(t)$ is maximum when $t = (\tau_F \tau_R \ln(\tau_R/\tau_F))/(\tau_R - \tau_F)$. The terms in Equation 3.1 can be expressed in the following forms (Messenger, 1982):

$$I_0 = q\,\mu\,N\,E_0\,\sec(\theta) \tag{3.3}$$

$$\tau_F = [\mu\,dE(X)/dX]^{-1} \tag{3.4}$$

where $q = 1.602 \times 10^{-19}$ C is the electron charge, μ is the average mobility (which depends on the electric field $E[X]$), N is the electron–hole pair linear density (cm^{-1}), $E_0 = E(0)$ is the electric field at $X = 0$, and $dE(X)/dX$ is the change rate of the electric field with respect to the position. The electron–hole pair linear density depends on the absolute linear energy transfer (LET) in units of megaelectronvolts per centimetre (Holbert, 2012):

$$N = LET/E_{gSi} \tag{3.5}$$

where $E_{gSi} = 3.6$ eV is the necessary energy to create an electron–hole pair in silicon. LET depends on the particle kinetic energy $E_{k\alpha}$. The relative LET for a material in MeV-cm^2/mg for a particle is defined as

$$LET_M = LET/\rho_M \tag{3.6}$$

where ρ_M is the material volumetric density. In silicon, $\rho_M = \rho_{Si} = 2329$ mg/cm^3. The relative silicon LET is also expressed in other units as picocoulombs per micrometre (1 pC/μm = 96.525 MeV-cm^2/mg) (Naseer, 2008).

Table 3.1 shows the principal sources of natural space radiation (Ecoffet, 2007). These radiations can reach the earth. Other radiation sources include the manufacturing materials used in the fabrication of integrated circuits (Wrobel et al., 2009).

In the literature, the faults due to particle impacts are known as single-event effects because they are due to a single particle or heavy ion. The single-event effects have subcategories (Boudenot, 2007), such as

- Single-event transient (SET): Transient fault in combinational circuits
- Single-event upset (SEU): Transient fault in sequential circuits and memories

TABLE 3.1

Main Sources of Natural Space Radiation

Radiation belts	Electrons	1 eV to 10 MeV
	Protons	1 keV to 500 MeV
Solar flares	Protons	1 keV to 500 MeV
	Ions	1 to few 10 MeV/n
Galactic cosmic rays	Protons and ions	Max flux at about 300 MeV/n

Source: Ecoffet, F., in Velazco, R., et al. (Eds.), *Radiations Effects on Embedded Systems*, Springer, Berlin, 2007, pp. 31–68.

- Single-event latch-up (SEL): Destructive fault; can affect the CMOS structure
- Single-event burnout (SEB): Destructive fault; affects power metal oxide semiconductor field-effect transistors (MOSFETs)
- Single-event gate rupture (SEGR): Fault that can damage the submicron structure
- Single hard error (SHE): Destructive fault in complex circuits

SET and SEU are the only transient faults. SEU is a failure that changes the value of a bit in a register or memory cell. A register with a logical value 1 is changed to a logical 0 after being affected or vice versa. SEU failures are also known as soft errors. Since the SET affects the functionality of transistors, creating an anomalous current, it can affect the final result of a logic operation. These transient faults can introduce a temporary error, so they will not affect future circuit operation.

3.3 CMOS Bulk and FDSOI Devices

The traditional industry standard, the MOSFET bulk technology, is facing problems with static power consumption and other second-order effects, in technology nodes below 130 nm. To try to handle these problems, one can explore new materials to replace silicon, such as hybrids like Ge-Si or gallium-arsenide, or try to replace the gate silicon oxide by other types of insulating materials to keep up with Moore's law. Other devices are being developed in 3D, such as Fin-FETs, to keep increasing transistor density.

A set of structures was designed using TCAD simulation tools. Figure 3.3a shows a 32 nm predictive technology model (PTM) n-type MOSFET (NMOS) bulk transistor. This transistor has a p-type substrate doping of 4.12×10^{18} cm^{-3}, a junction depth of 50 nm, a silicon oxide thickness t_{ox} of 1.3 nm and a metal-gate work function Φ_M of 4.25 eV. Figure 3.3b presents a 28 nm FDSOI high-K NMOS transistor. This transistor is a p-type one, and it has a substrate doping of 1×10^{14} cm^{-3}, a p-type channel doping of 1×10^{15} cm^{-3} with a thickness of 8.5 nm, an equivalent gate oxide thickness t_{EOX} of 0.75 nm (SiO$_2$ thickness of 0.55 nm and HfO$_2$ thickness of 1.283 nm), a buried oxide (BOX) thickness of 20 nm, a p-type back plane (BP) doping of 2×10^{18} cm^{-3} with a thickness of 25 nm and a metal-gate work function Φ_M of 4.52 eV. Another 28 nm FDSOI transistor was also created with a 0.9 nm silicon oxide thickness, with the same characteristics and metal-gate work function as the 28 nm FDSOI high-K, to compare other geometry effects. Figure 3.4 shows a comparison between the I_d and V_g curves of these devices

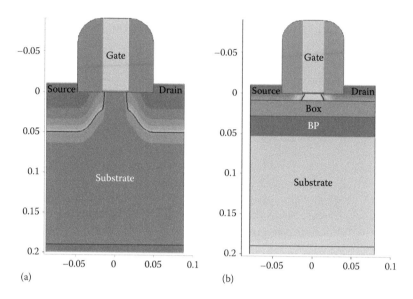

FIGURE 3.3
Devices designed using TCAD tools. (a) 32 nm NMOS bulk. (b) 28 nm NMOS FDSOI.

using the TCAD-created structures and the corresponding SPICE model card. For both devices, the width is $W_g = 300$ nm. Table 3.2 presents the characteristics of these devices under test.

3.4 Heavy-Ion Impact Simulation on Single Devices

In this case, the simulations were performed with the presented devices in the off state, such as shown in Figure 3.5a and b. Figure 3.5a presents a 32 nm bulk CMOS transistor setup, and Figure 3.5b shows a 28 nm FDSOI setup, where the BP terminal is grounded. In both cases, each device is biased with 1 V on the drain terminal. The device widths are $W_g = 100$ nm for both.

The heavy ion for the simulation was configured with LET = 100 MeV-cm²/mg (or 1.0447 pC/μm), a total track range $l = 300$ nm and a characteristic distance $w_t = 20$ nm. The total simulation time was $T_s = 100$ ps, and the heavy ion impacted at $t_i = 25$ ps (Calienes et al., 2015).

The simulated heavy ion impacted the raised terminals at six different angles θ (0°, 15°, 30°, 45°, 60° and 75°) and at five different locations L_i (measured in nanometres from transistor spacers: 6, 12, 18, 24 and 30 nm) for each angle; that means 30 simulations. Figure 3.5c presents how the heavy-ion impact is performed using the distance L_i and impact angle θ.

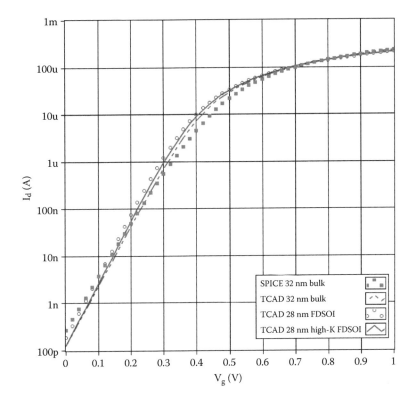

FIGURE 3.4
I_d vs. V_g calibration curves between SPICE and TCAD for 32 nm bulk, 28 nm FDSOI and 28 nm FDSOI high-K ($W_g = 300$ nm).

TABLE 3.2

Electrical Characteristics of the Studied Devices

Parameter Name	32 nm Bulk	28 nm FDSOI High-K	28 nm FDSOI
Turn-on current (I_{on})	778 µA/µm	727 µA/µm	702 µA/µm
Turn-off current (I_{off})	423 pA/µm	421 pA/µm	461 pA/µm
Saturation threshold voltage ($V_{th,SAT}$)	229 mV	219 mV	218 mV
Linear threshold voltage ($V_{th,LIN}$)	300 mV	252 mV	254 mV
Drain-induced Barrier Lowering (DIBL)	81 mV/V	38 mV/V	41 mV/V
Subthreshold slope (SS)	78 mV/dec	75 mV/dec	75 mV/dec

The heavy-ion impacts on the drain and source terminals produce a transient current. To obtain the collected charge (CC) after the ion hit, it is necessary to integrate the transient current with respect to time for each L_i-θ pair (Calienes et al., 2015).

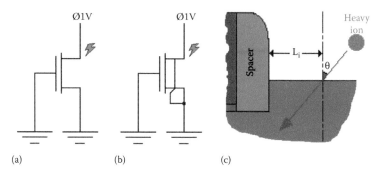

FIGURE 3.5
Simulation setup. (a) 32 nm bulk. (b) 28 nm FDSOI. (c) Heavy-ion impact setup. The "ø" symbol is part of circuits scheme in (a) and (b).

3.4.1 Bulk Transistor of 32 nm

3.4.1.1 Heavy-Ion Impacts the Drain Terminal

The heavy ion with LET = 100 MeV-cm^2/mg was impacted on the raised drain terminal of the 32 nm bulk transistor using the circuit configuration presented in Figure 3.5a, and the resulting CC is presented in Figure 3.6. The heavy ion produces a maximum CC of 32.29 fC when L_i = 30 nm and θ = 30°. The minimum CC is 13.12 fC when L_i = 6 nm and θ = 75°. The tendency in these conditions is for CC to decrease when the impact is near to the nitride spacer and the angle is increased.

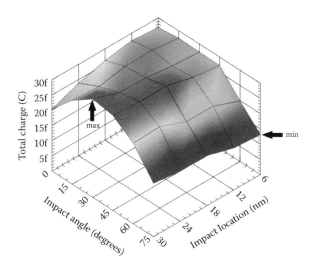

FIGURE 3.6
Total CC results of 100 MeV-cm^2/mg heavy-ion impact on 32 nm bulk transistor drain terminal (Φ_M = 4.25 eV, W_g = 100 nm). (From Calienes et al., 2015.)

When the L_i value is increased, CC tends to stay constant, with little charge variation; for example, with $\theta = 45°$, CC is 27.27 fC for $L_i = 6$ nm, 30.80 fC for $L_i = 18$ nm and 31.27 fC for $L_i = 30$ nm. When $\theta = 15°$, CC is even more constant.

When the impact angle θ is increased, the CC variation presents a 'sinusoidal' behaviour in all impact locations L_i; that is, CC is low at $\theta = 0°$, higher at $\theta = 30°$ and at $\theta = 75°$ is lower than the charge at $\theta = 0°$, as Figure 3.6 presents.

3.4.1.2 Heavy-Ion Impacts the Source Terminal

When the heavy ion impacted the source terminal of a 32 nm bulk transistor in the off state, the CC was less than the drain terminal impact case. Figure 3.7 shows this CC behaviour. The ion produces a maximum CC of 29.89 fC when $L_i = 12$ nm and $\theta = 60°$, and the minimum CC is 2.31 fC when $L_i = 30$ nm and $\theta = 0°$.

When L_i is increased, the CC tends to decrease slowly, almost constantly; for example, when $\theta = 30°$, for $L_i = 6$ nm the CC is 18.04 fC, for $L_i = 18$ nm it is 15.45 fC and for $L_i = 30$ nm it is 12.99 fC, as Figure 3.7 shows.

In the case of an increase in θ, the CC increases up to $\theta = 60°$, and then it declines slightly when $\theta > 60°$; for example, when $L_i = 24$ nm, for $\theta = 0°$ the

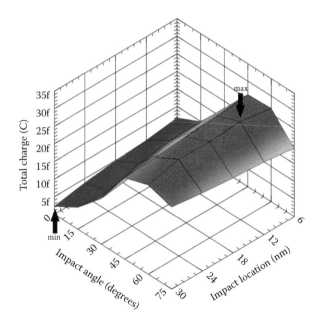

FIGURE 3.7
Total CC results of 100 MeV-cm²/mg heavy-ion impact on 32 nm bulk transistor source terminal ($\Phi_M = 4.25$ eV, $W_g = 100$ nm).

CC is 3.83 fC, for $\theta = 30°$ it is 14.24 fC, for $\theta = 60°$ it is 27.49 fC and for $\theta = 75°$ it is 22.25 fC.

3.4.2 High-K FDSOI Transistor of 28 nm

3.4.2.1 Heavy-Ion Impacts the Drain Terminal

In this case, the 100 MeV-cm^2/mg heavy ion impacted the 28 nm high-K FDSOI drain terminal. The simulation was conducted in the same way as the simulation of the bulk transistor. Figure 3.8 shows the CC as a function of L_i and $\theta°$. In this case, 5.13 fC is the maximum CC and it occurs when $L_i = 24$ nm and $\theta = 75°$. The minimum CC is 0.45 fC when $L_i = 30$ nm and $\theta = 0°$, a vertical impact far from the nitride spacer.

When L_i is increased, the CC tends to decrease; for example, for $\theta = 45°$, at $L_i = 6$ nm the charge is 2.78 fC, at $L_i = 18$ nm it is 2.23 fC and at $L_i = 30$ nm it is 1.31 fC. There are several exceptions at $\theta = 30°$ and $\theta = 60°$, but in general, the trend is met. If θ increases, the CC also increases. The minimum CC is when $\theta = 0°$ and the maximum when $\theta = 75°$, regardless of the impact location L_i.

3.4.2.2 Heavy-Ion Impacts the Source Terminal

When the 100 MeV-cm^2/mg heavy-ion impacts the 28 nm high-K FDSOI source terminal, the maximum CC value is 4.46 fC at $L_i = 6$ nm and $\theta = 75°$,

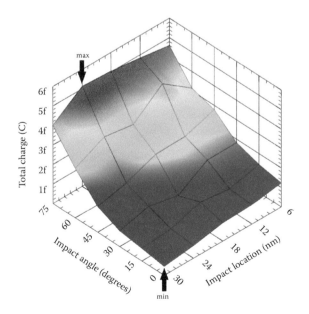

FIGURE 3.8
Total CC results of 100 MeV-cm^2/mg heavy-ion impact on 28 nm high-K FDSOI transistor drain terminal ($\Phi_M = 4.52$ eV, $W_g = 100$ nm).

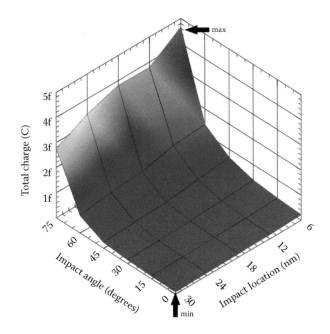

FIGURE 3.9
Total CC results of 100 MeV-cm^2/mg heavy-ion impact on 28 nm high-K FDSOI transistor source terminal ($\Phi_M = 4.52$ eV, $W_g = 100$ nm).

very close to the nitride spacer, and the minimum is 0.096 fC at $L_i = 30$ nm and $\theta = 0°$, far from the nitride spacer. Figure 3.9 presents these CC tendencies in function of L_i and $\theta°$.

When L_i increases, the CC tends to decrease slowly; for example, for $\theta = 30°$, at $L_i = 6$ nm the CC is 0.38 fC, at $L_i = 18$ nm it is 0.24 fC and at $L_i = 30$ nm it is 0.15 fC. The exception is at $\theta = 75°$, when the CC variation in function of L_i is greater than at other impact angles. When θ is increased, the CC tends to increase in all cases. Regardless of the L_i value, the maximum CC occurs when $\theta = 75°$, and the minimum when $\theta = 0°$.

3.4.3 FDSOI Transistor of 28 nm

3.4.3.1 Heavy-Ion Impacts the Drain Terminal

The 100 MeV-cm^2/mg heavy-ion impact on the 28 nm FDSOI drain terminal was performed in the same way as in the previous simulations. Figure 3.10 presents the result of the TCAD simulation for this case. The maximum CC is 4.20 fC and occurs when $L_i = 12$ nm and $\theta = 75°$, close to the drain nitride spacer. The minimum CC for this case is 0.41 fC at $L_i = 30$ nm and $\theta = 0°$, vertical and far from the spacer.

When L_i increases, the CC tends to decrease; for example, for $\theta = 60°$, at $L_i = 6$ nm the CC is 3.69 fC, at $L_i = 18$ nm the charge is 3.50 fC and at

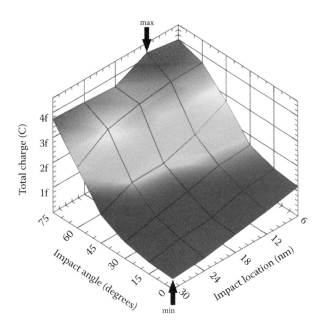

FIGURE 3.10

Total CC results of 100 MeV-cm²/mg heavy-ion impact on 28 nm FDSOI transistor drain terminal (Φ_M = 4.52 eV, W_g = 100 nm).

L_i = 30 nm it is 2.42 fC. The exception is when the impact occurs with θ = 75°. In this case, the charge increases slowly from 3.58 fC at L_i = 18 nm to 3.74 fC at L_i = 30 nm. When the angle θ increases, the CC also increases in all cases. The minimum CC values occur at θ = 0° and the maximum values occur when θ = 75°, regardless of the impact location L_i.

3.4.3.2 Heavy-Ion Impacts the Source Terminal

For the 28 nm FDSOI source terminal impact case, the maximum CC is 4.10 fC when L_i = 6 nm and θ = 75°, and the minimum charge is 0.0706 fC at L_i = 30 nm and θ = 0°. The CC behavior in this case is similar to the one with a drain impact, but in this case the CC is much lower. These results are shown in Figure 3.11.

3.4.4 Conclusions Related to Heavy-Ion Impact Simulation on Single Devices in Different Technologies

The collected transient charge and drain current peak depend on the substrate bias, lightly doped drain (LDD) geometry, silicon volume in the body/channel region, gate equivalent oxide thickness and polymeric metal-gate materials (Calienes et al., 2015). The generated charge due to an ion impact is collected by recombination, drift and diffusion processes. The

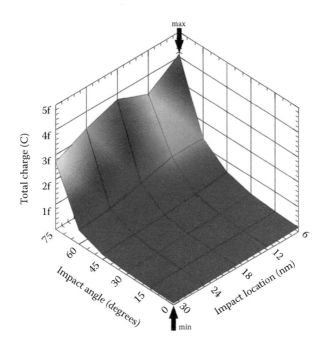

FIGURE 3.11
Total CC results of 100 MeV-cm²/mg heavy-ion impact on 28 nm FDSOI transistor source terminal ($\Phi_M = 4.52$ eV, $W_g = 100$ nm).

transient peak current depends directly on the drift current component and substrate bias.

The quantity of CC is directly proportional to the silicon volume, and it also depends on the doping of the body (Calienes et al., 2015). In the ultrathin body and box (UTBB) FDSOI case, the silicon body thickness is less than the channel length and limited by the BOX. The recombination is lower than in the bulk transistor case, because the FDSOI silicon body is much thinner than in the bulk one. In the worst case, the CC in the FDSOI simulated transistors is smaller by a factor of approximately 7.68 than the worst-case CC in a bulk transistor. The electron density in the device is a measure of how the heavy-ion impact affects a device. Figures 3.12 through 3.14 show the variation over time of the electron density for each simulated device in the worst CC cases when a 100 MeV-cm²/mg heavy-ion impacts the drain terminal.

The most sensitive device area is the reverse-biased drain and source n-p junction (Messenger, 1982). In a bulk device, the maximum CC occurs when $L_i = 30$ nm and $\theta = 30°$, when the ion track funnels through the LDD around the drain region, as shown in Figure 3.15a. A similar phenomenon occurs with a FDSOI device when $L_i = 12$ nm and $\theta = 75°$, as in Figure 3.15b. In the high-K FDSOI case, this maximum CC occurs when $L_i = 24$ nm and $\theta = 75°$. The CC difference between FDSOI and high-K FDSOI is small: 0.93 fC. In all cases, the ion funnels through the LDD and produces a maximum CC. Also,

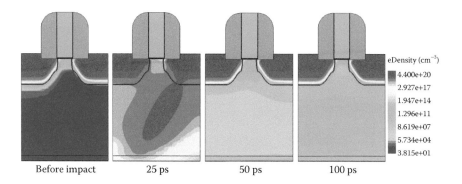

FIGURE 3.12
Electron density over time in 32 nm bulk transistor when Li = 30 nm and θ = 30° (Φ_M = 4.25 eV, W_g = 100 nm).

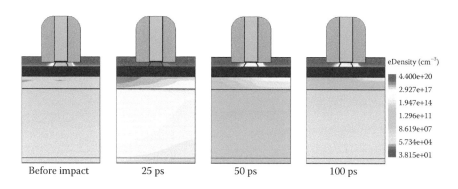

FIGURE 3.13
Electron density over time in 28 nm FDSOI transistor when Li = 12 nm and θ = 75° (Φ_M = 4.52 eV, W_g = 100 nm).

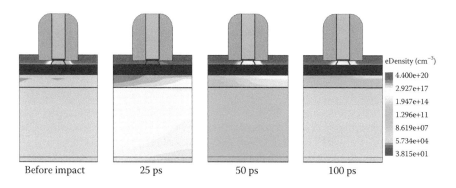

FIGURE 3.14
Electron density over time in 28 nm FDSOI high-K transistor when Li = 24 nm and θ = 75° (Φ_M = 4.52 eV, W_g = 100 nm).

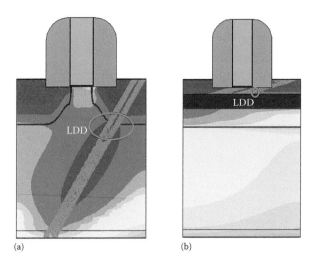

FIGURE 3.15
LDD region and heavy-ion funnelling. (a) 32 nm bulk (Li = 30 nm, θ = 30°). (b) 28 nm FDSOI
(Li = 12 nm, θ = 75°).

the LDD doping in the bulk transistor is $-8 \times 10^{17} \mathrm{cm}^{-3}$ and the LDD doping
in both FDSOI transistors is approximately $-2 \times 10^{14} \mathrm{cm}^{-3}$. The silicon doping
in the body also has influence on the CC.

Simulations were also performed using different metal-gate work function
Φ_M values in all transistors. Also, in the FDSOI cases, simulations were
performed using devices with different gate equivalent oxide thickness t_{EOX}.
The materials of the polymeric metal gate of the devices and the gate oxide
thickness have influence on CC. The CC is formed by mobile charges generated when the heavy-ion funnels through the device, and the charge of
the depletion zone Q_d. The charge Q_d depends on the Φ_M and gate oxide
capacitance $C_{ox} = \varepsilon_{EOX}/t_{EOX}$:

$$Q_d \sim C_{ox}(Vth - \Phi_M + \Phi_S + q\left(N_{tox}/C_{ox}\right) - 2\phi_F) \quad (3.7)$$

where Φ_S is the body/channel semiconductor work function, $q = 1.602 \times 10^{-19} \mathrm{C}$
is the electron charge, N_{tox} is the total oxide charge surface density and ϕ_F
is the silicon Fermi potential. So, increasing Φ_M and t_{EOX} decreases Q_d, and
also decreases the total SET CCs (Calienes et al., 2015). If t_{EOX} decreases, the
device is predisposed to produce more CC due to a heavy-ion impact. The
CC can be modelled as the sum of Q_d and the mobile charges produced by
the heavy-ion impact. Figure 3.16 presents a comparison between two 28 nm
FDSOI transistors ($t_{ox} = 0.9$ nm) with different Φ_M: 4.25 and 4.52 eV (these
data are used to obtain the data of Figure 3.10). The heavy ion has the same
LET = 100 MeV-cm²/mg and characteristics as in the previous simulations.
The simulated impacts occur in two L_i locations: 12 and 30 nm. Therefore, the

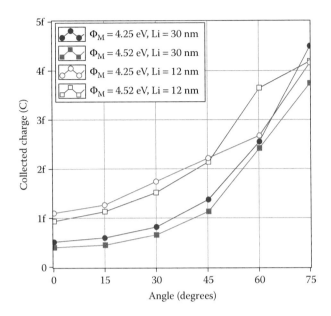

FIGURE 3.16
Comparison between two 28 nm FDSOI transistors with different Φ_M and two different drain impact locations L_i.

transistor with $\Phi_M = 4.25$ eV collects more charge due to a heavy-ion impact than the $\Phi_M = 4.52$ eV one in a linear proportional relation, almost as predicted in Equation 3.7. This relation is no longer valid when the ion funnels through the drain, the LDD region and the body/channel at the same time. Figure 3.16 shows what happens in this case, when $\theta = 60°$ and $L_i = 12$ nm: the CC in the $\Phi_M = 4.52$ eV case is greater than when $\Phi_M = 4.25$ eV.

The drain terminal is the most sensitive area in these devices, because in all cases, the heavy-ion impact on this terminal produces more CC than the impact on the source terminal. In this case, $V_{ds} = V_{db} = 1$ V is the voltage at the drain terminal and $V_{sb} = 0$ V is the voltage between the source and substrate terminals. In this case, the drain-body n-p junction is reverse biased ($V_{db} = 1$ V and the quasi-Fermi levels of the drain and body are different) and there is a large depletion region with a lot of depletion charge. In the source-body n-p junction case, the depletion region is thin due to the very low-voltage difference (in this case, $V_{sb} = 0$ V and the quasi-Fermi levels are almost the same). These depletion charges increase the total CC when the ion funnels through the device. In the bulk transistor, it is possible to have almost the same CC if the ion impacts both terminals with an angle $\theta > 45°$, because there is the possibility to funnel into the LDD region from both the drain and source impact locations. In the bulk transistor, the channel region has a relativity high doping, and this increases the final SET CC. For the FDSOI devices, the low-doped channel region has a low charge contribution to CC after a heavy-ion impact. This fact, combined with the thin body/channel region and the bigger BOX region below the body/channel in the FDSOI transistor, would necessitate a very sharp angle ($\theta > 70°$) for the ion to funnel through the LDD region and to obtain almost the same CC as having a heavy-ion impact on the source or drain terminals.

The BP and BOX in the FDSOI devices have advantages in a radiation environment. When the heavy-ion funnels through the device, the charges do not return to the channel, because the BOX isolates the charges, avoiding a contribution to the drain current. The grounded BP discharges the transistor substrate slowly (Calienes et al., 2015).

3.5 Heavy-Ion Impact Simulation on 6T Static RAM Cells

In order to simulate the effect of a heavy-ion impact on a six-transistor SRAM (6T SRAM) in these three devices, with the goal of comparing the radiation effects, it is necessary to create and set up the circuits for the test. The objective is to see what happens with the data stored in the cell and what is the minimum particle LET and CC to produce an SEU. The simulated memory cell schematic is shown in Figure 3.17 (Calienes et al., 2014). The cell is in retention mode with a supply voltage of 1 V, the word line (WL) is at 0 V and the bit lines (BL and BL') have 0 A current supply sources, to simulate high impedance. In this case, the TCAD simulation was performed in mixed mode; that is, five transistors were described using a SPICE model card and the impacted transistor was described at the device level. Tables 3.3 and 3.4 present the dimensions of each transistor in the three test circuits. The mnl transistor is the one suffering a heavy-ion impact in the drain terminal in the

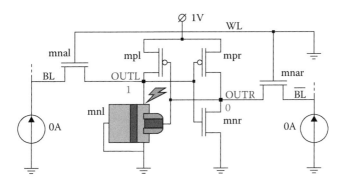

FIGURE 3.17
Mixed-mode 6T SRAM cell. (From Calienes, W., et al., Impact of SEU on bulk and FDSOI CMOS SRAM, presented at Proceedings of 10th Workshop of the Thematic Network on Silicon-on-Insulator Technology, Devices and Circuits, Tarragona, Spain, January 29, 2014.)

TABLE 3.3

Transistor Dimensions of 32 nm Bulk 6T SRAM Cell

Transistor	Transistor Width W (nm)	Terminal Perimeter P (nm)	Terminal Area A (nm²)	W/L
mnal	160	560	19,200	5
mnar	160	560	19,200	5
mnr	217	674	26,040	6.78
mpr	91	422	10,920	2.84
mnl	217	674	26,040	6.72
mpl	91	422	10,920	2.84

TABLE 3.4

Transistor Dimensions of 28 nm FDSOI and 28 nm High-K FDSOI Memory Cells

Transistor	Transistor Width W (nm)	Terminal Perimeter P (nm)	Terminal Area A (nm²)	W/L
mnal	140	520	16,800	5
mnar	140	520	16,800	5
mnr	190	620	22,800	6.79
mpr	80	400	9,600	2.86
mnl	190	620	22,800	6.79
mpl	80	400	9,600	2.86

worst-case CC for each device type. Transistors mnl and mnr have a bigger area than the other ones in this circuit. To simulate a logical 1 in the cell, the mnl drain (the OUTL node) is initialised at 1 V, while the mnr drain (the OUTR node) is initialised at 0 V. The BPs of the 28 nm FDSOI and 28 nm high-K FDSOI cells are tied to ground.

The total simulation time in all cases is $T_s = 1$ ns, and the heavy-ion impact occurs at $t_i = 0.490$ ns. The LET is variable. The Li and θ values are chosen from the previous device simulations to obtain the maximum CC.

3.5.1 Simulation Results of 32 nm Bulk 6T SRAM

Heavy ions with 1–10 MeV-cm²/mg LETs were used to strike the drain terminal at $L_i = 30$ nm and $\theta = 30°$ (the most disruptive condition for heavy-ion impact when the transistor is off state). Figure 3.18 shows the behaviour of the CC in the transistor mnl with different LETs. In these simulations, the minimum LET to flip the memory cell is 5 MeV-cm²/mg, and it produced 1.91 fC of CC, but the minimum CC to flip the cell (the 'critical charge') for this circuit was $C_{ch} = 1.76$ fC. Using this C_{ch} (Figure 3.18), the minimum estimated LET to flip the memory cell is approximately 4.75 MeV-cm²/mg. The grey area in Figure 3.18 is the critical area where the memory cell has a bit flip.

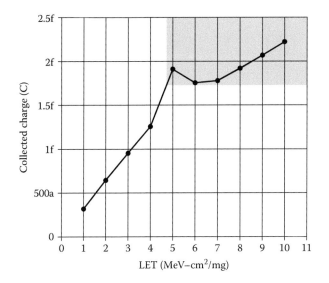

FIGURE 3.18
CC vs. LET characteristics of 32 nm bulk 6T SRAM cell (L_i = 30 nm, θ = 30°). The grey area is the critical area when the cell flips.

3.5.2 Simulation Results of 28 nm FDSOI 6T SRAM

In this case, heavy ions with 60–70 MeV-cm^2/mg LETs were used to impact the mnl transistor drain terminal at L_i = 12 nm and θ = 75° (the worst case for this device in previous simulations). Figure 3.19 presents the results of the simulations for this memory cell. The grey area represents the critical area for the circuit. In the simulations, the cell flipped with LET = 64 MeV-cm^2/mg and produced 2.03 fC of CC, but the C_{ch} in this case was 1.78 fC when LET = 70 MeV-cm^2/mg. The estimated LET to produce the C_{ch} was approximately 63.6 MeV-cm^2/mg, but it is possible to improve this result with more simulations using LET up to 90 MeV-cm^2/mg. Figure 3.20 presents the voltage variation in the OUTL and OUTR nodes when a 64 MeV-cm^2/mg heavy-ion impacts the mnl transistor drain terminal at L_i = 12 nm and θ = 75°.

3.5.3 Simulation Results of 28 nm FDSOI High-K 6T SRAM

In this simulation, the heavy-ion impacts on the mnl transistor were performed using 45–55 MeV-cm^2/mg LETs. The drain terminal was hit at L_i = 24 nm and θ = 75°. Figure 3.21 shows the simulation results. The critical area for the circuit is shaded in grey. The minimum LET when the memory cell flipped was 50 MeV-cm^2/mg, and it produced a CC of 1.89 fC. The critical charge is approximately C_{ch} = 1.80 fC when LET = 55 MeV-cm^2/mg. LET = 49.7 MeV-cm^2/mg is the minimum estimated LET to generate this approximate C_{ch}.

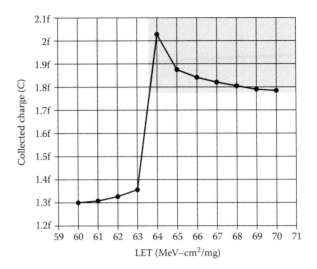

FIGURE 3.19
CC vs. LET characteristics of 28 nm FDSOI 6T SRAM cell ($L_i = 12$ nm, $\theta = 75°$). The grey area is the critical area when the cell flips.

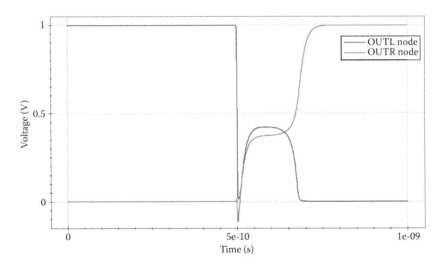

FIGURE 3.20
OUTL and OUTR nodes voltages when a 64 MeV-cm^2/mg heavy-ion impacts the mnl transistor drain terminal in a 28 nm FDSOI memory cell.

3.5.4 Conclusions Related to the Heavy-Ion Impact Simulation on 6T Static RAM Cells

The flipping of a memory cell depends on the LET of the particle and the impact location in the transistor. In this case, to study the radiation effects,

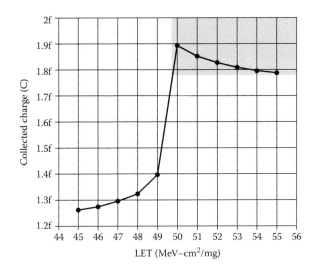

FIGURE 3.21
CC vs. LET characteristics of 28 nm FDSOI high-K 6T SRAM cell ($L_i = 24$ nm, $\theta = 75°$). The grey area is the critical area when the cell flips.

only impacts on the large NMOS transistors mnl and mnr were considered, but this methodology is also valid in the case of p-type MOSFET (PMOS) transistors, when the memory cell is in retention mode. All measurements were done using the most critical locations and angles, that is, where the device is more susceptible to produce more CCs.

The 28 nm FDSOI memory cell is more resilient against heavy-ion impacts than the 32 nm bulk and 28 nm FDSOI high-K ones. The comparison of transient current pulses due to the heavy-ion impact, considering the worst cases for each cell, is shown in Figure 3.22. Table 3.5 shows the comparison among these cases. The 28 nm FDSOI cell is (in LET terms) 12.8 times more resilient than the 32 nm bulk cell, and 1.28 times more resilient than the 28 nm FDSOI high-K cell. In the three cases, the CC that flips a cell is almost the same, nearly 2 fC. The estimated critical charge C_{ch} is almost the same in the three cases: 1.78 fC. A CC $\geq C_{ch}$ is needed to flip these memory cells. Therefore, almost the same C_{ch} needs to be generated to flip a cell in the three cases.

Figure 3.23 shows the electron density distribution as a function of time in the impacted transistor, in a 28 nm FDSOI high-K memory cell. A 50 MeV-cm^2/mg heavy-ion impacts the drain terminal at $T_S = t_i = 0.5$ ns. The ion produces CCs, and a transient current pulse is generated as shown in Figure 3.22, and the cell is flipped. The probability of the generation of a current pulse also depends on how the transistors are connected in the circuit and on the circuit layout. When the ion impacts the mnl transistor drain terminal, the device is turned on briefly due to the charges produced by the impact; this voltage-level change in OUTL is sufficient to set in motion a positive feedback and flip the right-side inverter. This inverter changes its

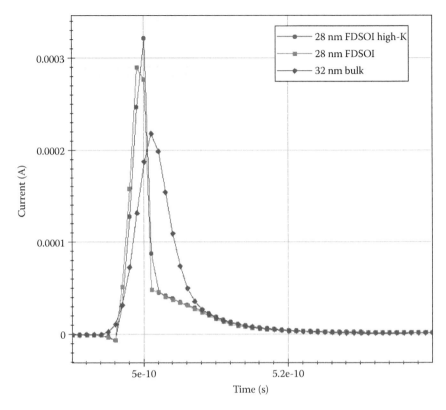

FIGURE 3.22
mnl transistor drain current for each simulated memory cell.

TABLE 3.5

Comparison of the Studied Devices in a 6T SRAM

Memory Cell Type	LET (MeV-cm²/mg)	CC (fC)	C_{ch} (fC)	Maximum Pulse Amplitude (µA)
32 nm bulk	5	1.91	1.76	218
28 nm FDSOI	64	2.03	1.78	290
28 nm FDSOI high-K	50	1.90	1.80	321

output OUTR value to an effective 'high level' and turns on the mnl transistor (last picture of Figure 3.23).

The difference of the minimum LET to flip the 32 nm bulk memory cell and the 28 nm FDSOI memory cells can be explained by the difference in silicon volume of the affected device in each memory cell. The generated CC in the silicon is directly proportional to the LET times the distance travelled by the particle (Calienes et al., 2014).

The difference of minimum LETs between the 28 nm FDSOI memory cell and the 28 nm FDSOI high-K memory cell has another explanation. When the

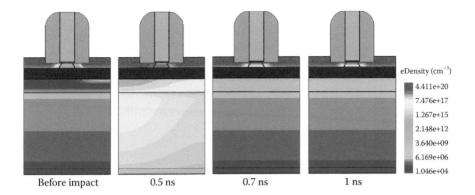

FIGURE 3.23
Electron density time variation of 28 nm FDSOI high-K memory cell mnl transistor ($L_i = 24$ nm, $\theta = 75°$, LET = 50 MeV-cm²/mg).

equivalent oxide thickness t_{EOX} is thinner, the C_{ox} is greater, and the depleted charge Q_d is directly proportional to C_{ox}, as shown in Equation 3.7. In the 28 nm FDSOI, $t_{ox} = 0.9$ nm, and this is the minimum silicon oxide thickness to avoid the gate tunnel effect. In the 28 nm FDSOI high-K, $t_{ox} = t_{EOX} = 0.75$ nm for the polymeric HfO_2-SiO_2 gate oxide. This silicon oxide thickness of the high-K cell avoids the gate tunnel effect because the total sum of the two oxide layers is 1.833 nm, which is greater than 0.9 nm. This is one of the reasons to choose high-K devices when the technology node is less than 90 nm.

3.6 Conclusions

The main conclusion is that a FDSOI CMOS transistor is more resilient to heavy-ion impacts than a bulk CMOS one. The volume of silicon in the active region and its doping are deciding characteristics in choosing this technology to cope with heavy-ion impacts. The BOX and BP regions avoid the back recombination in the channel region, depriving any drain current from the substrate charge. The CC with the FDSOI transistor was approximately 7.7 times less than with an identically sized bulk transistor in the worst case.

In all cases, the drain terminal was more sensitive to heavy-ion impacts than the source terminal. The CC depended on the depletion region thickness in the impacted terminal.

The SET CC in the tested devices did not depend on the impact angle. The CC depended on silicon volume in the body/channel, the body/channel doping, the gate equivalent oxide thickness, the polymeric metal-gate work function, the source/drain contact diffusion geometry and the LDD location.

The 28 nm FDSOI memory cells were more resilient than the 32 nm bulk ones. It can be concluded that for the three types of cells, the same amount of charge needs to be generated in the silicon to flip the SRAM cells. FDSOI also has the advantage of using a much thinner layer in the body/channel region.

The 28 nm FDSOI memory cell was more resilient than the 28 nm FDSOI high-K one. To avoid the tunnel effect in the gate insulator, it is necessary to choose the high-K cell. An equivalent oxide with a thin thickness predisposes a device to collect more charges when a heavy-ion impacts the device.

References

Boudenot, J. C. 2007. Radiation space environment. In Velazco, R., Fouillat, P., and Reis, R. (Eds.), *Radiations Effects on Embedded Systems*, Springer, Berlin, pp. 1–9.

Calienes, W., and Reis, R. 2011. SET and SEU simulation toolkit for LabVIEW. Presented at Proceedings of European Conference on Radiation and Its Effects on Components and Systems (RADECS), Seville, Spain, September 19, 2011.

Calienes, W., Reis, R., Anghel, C., and Vladimirescu, A. 2014. Impact of SEU on bulk and FDSOI CMOS SRAM. Presented at Proceedings of 10th Workshop of the Thematic Network on Silicon on Insulator Technology, Devices and Circuits, Tarragona, Spain, January 29, 2014.

Calienes, W., Vladimirescu, A., and Reis, R. 2015. Bulk and FDSOI sub-micron CMOS transistors resilience to single-event transients. Presented at Proceedings of 2015 IEEE International Conference on Electronics, Circuits, and Systems (ICECS 2015), Cairo, December 6–9, 2015.

Ecoffet, F. 2007. In-flight anomalies on electronic devices. In Velazco, R., Fouillat, P., and Reis, R. (Eds.), *Radiations Effects on Embedded Systems*, Springer, Berlin, pp. 31–68.

Holbert, K. 2012. Charged particle ionization and range. Available at http://holbert.faculty.asu.edu/eee460/IonizationRange.pdf.

Messenger, G. C. 1982. Collection of charge in junction nodes from ion tracks. *IEEE Transactions of Nuclear Science*, vol. NS-26, no. 6, pp. 2024–2031, 1982.

Munteanu, D., and Autran, J.-L. 2008. Modeling and simulation of single-event effects in digital devices and ICs. *IEEE Transactions on Nuclear Science*, vol. 55, no. 4, pp. 1854–1878, 2008.

Naseer, R. 2008. A framework for soft error tolerant SRAM design. Thesis, University of Southern California, Los Angeles.

Wrobel, F., Saigne, F., Gedion, M., Gasiot, J., and Schrimpf, R. D. 2009. Radioactive nuclei induced soft errors at ground level. *IEEE Transactions on Nuclear Science*, vol. 56, no. 6, pp. 3437–3441, 2009.

Section II

Sensors and Operating Conditions

4

Electronic Sensors for the Detection of Ovarian Cancer

A. M. Whited and Raj Solanki

CONTENTS

ABSTRACT   Ovarian cancer is a complex and multifaceted disease with a disproportionately high mortality rate, as it is most commonly diagnosed in the late stages of the disease, when the cancer has metastasised to surrounding tissues. Early detection of ovarian cancer–associated biomarkers CA-125, He4, and carcinoembryonic antigen (CEA) would be instrumental not just in detecting ovarian cancer while it is still in its early stages, but also in monitoring the effectiveness of treatment. The emergence of electrical biosensors holds the greatest promise in fulfilling these needs. Electrical biosensors integrate a sensing element and a signal transducer into one convenient device, decreasing the cost and time required for traditional laboratory testing. The portability of biosensors also makes them ideal in point-of-care situations and remote locations. Much work has been put into developing biosensors for CA-125, He4 and CEA, but for a true improvement in this aspect of women's health, a sensor must be developed that is capable of detecting all three of these biomarkers simultaneously. A rational approach to the design of such a sensor is described.

4.1 Need for Ovarian Cancer Biosensors

Although it is not a commonly diagnosed disease, being just the eighth most common cancer diagnosed among American women each year, ovarian cancer is the fifth leading cause of cancer death in women [1]. Due to its nonspecific symptoms, which include bloating, a distended abdomen and gastrointestinal difficulties [2], and the absence of an adequate physical exam, more than 75% of cases are diagnosed in the later stages when the disease has metastasised beyond the primary site, leading to a 5-year survival rate of less than 30% [3]. By comparison, patients diagnosed in the early stages have a 90% survival rate over the same period [3]. These statistics are evidence that the development of an accurate and sensitive screening method capable of early detection of ovarian cancer would save thousands of women's lives each year.

There are three types of ovarian cancer: epithelial (90% of cases), germ cell carcinoma (5% of cases) and stromal carcinoma (5% of cases). An ovarian cancer diagnosis results in the classification of the disease into stages I–IV, with the lower-numbered stages being used to diagnose the early state of the disease and the stage of the disease associated with the highest survival rates. In stages III and IV, the cancer has metastasised beyond the primary site into surrounding tissue, resulting in a poorer prognosis. Unfortunately, a majority of diagnoses occur in these late stages, when the survival rate is below 30% [3].

Most diseases have known associated biomarkers, and ovarian cancer is no exception. CA-125 and He4 are the two protein biomarkers with the highest degree of correlation to ovarian cancer [4]. Currently, studies are under way to determine if the combined presence of CA-125 and He4 can be used to determine the presence of early stage ovarian cancer, making the combination a potentially powerful tool in disease diagnosis. As of 2008, both proteins are approved by the Food and Drug Administration (FDA) for use in monitoring for treatment efficacy and disease recurrence, but more work still remains to be done to determine their usefulness in early disease detection. While neither of these markers may be suitable for early detection of ovarian cancer on their own due to variations in expression level based on the type of malignancy and the presence of other benign gynaecological conditions, an elevated presence of both, coupled with a symptom index, would appear to make early diagnosis of ovarian cancer more feasible [5].

CA-125 has long been the gold standard biomarker in ovarian cancer diagnosis and in monitoring disease recurrence. It is still a poorly described protein despite being discovered more than 30 years ago [6]. Its structure and function remain largely a mystery, and it was only successfully cloned in 2001, despite the ability of research groups and biological reagent suppliers to produce the antibodies to it for some time [7]. It is a very large transmembrane protein weighing between 200 and 2000 kDa. It is composed of a short cytoplasmic tail, a transmembrane domain and a very large glycosylated extracellular domain, all of which are poorly characterised. CA-125 is not present in all types of ovarian cancer and can be elevated in a number of benign gynaecological conditions, such as endometriosis.

He4 is a more recently discovered ovarian cancer biomarker with much promise. It was discovered in 2003 and has been rapidly gaining ground as a robust biomarker for ovarian cancer [4]. It is found to be elevated in most patients at the early stage of the disease, and it is indicative of whether the disease has spread beyond the primary site [4]. For the time being, populations of healthy and diseased women are monitored to determine what constitutes an elevated level of He4. Despite its relatively recent discovery, He4 is less of a mystery than CA-125. The structure of He4 is fairly well known. It weighs 13 kDa, but in its usual glycosylated state, it weighs between 20 and 25 kDa [8]. Thus, it is a much smaller protein than CA-125. He4 is part of a family of whey acidic four-disulphide core (WFDC) proteins that are characterised by about 50 amino acid residues, including 8 conserved cysteine residues that form 4 disulphide bridges [9]. It is composed of one peptide and two of these WFDC regions [9]. He4 may be poised to become the signature biomarker for ovarian cancer [10], supplanting CA-125, as it appears to be more tightly relegated to just diseased tissue, thus holding promise for fewer false positives and, as a result, fewer surgeries, which come at a great cost to patients.

There are other biomarkers associated with ovarian cancer, although not quite as strongly as CA-125 and He4. One of them is carcinoembryonic antigen (CEA). CEA is most closely associated with colon cancer, being used to test for disease recurrence, but it has been reported in epithelial ovarian cancers, which are the most common type [11]. Like CA-125, it is a large transmembrane protein, weighing about 200 kDa. Although certainly not suitable for ovarian cancer diagnosis by itself, CEA has the promise to be a component of the ovarian cancer biomarker signature required for an accurate diagnosis.

Traditional bioassays to determine the presence of these three biomarkers in human blood samples are costly and time-consuming. A blood draw is required followed by lab analysis using enzyme-linked immunosorbent assay (ELISA). ELISA is a multistep process that relies on binding between an analyte, in this case CA-125, He4 or CEA, and an immobilised capture protein coated onto a solid substrate, usually a 96-well plate. Once the analyte has bound to the immobilised capture protein, a secondary, enzyme-conjugated protein is bound to the analyte. From there, the enzyme is used to create a reaction that generates a readable signal, generally a colour change. This diagnostic process requires specialised laboratory facilities and trained personnel, both of which add to the cost and time of disease diagnosis.

Biosensors have the potential to revolutionise the health-care industry and the way we diagnose diseases. A commercially viable biosensor that is used in a point-of-care environment has to be inexpensive, small, portable and capable of self-contained, rapid detection. Biosensors successfully integrate a sensing element and a signal transducer into one convenient platform. In addition to disease diagnosis, biosensors are useful in rapid follow-up care and their ability to continuously track disease progression and the effectiveness of treatment. Making them further ideal for disease detection, biosensors can be multiplexed, allowing the same device to detect more than one target molecule simultaneously. This is a necessary feature for diseases as complex as ovarian cancer, which we have already seen produces numerous associated biomarkers that are not always specific to the pathology being diagnosed or even expressed in all forms of the disease. Thus, the multiplexing feature provides an opportunity for highly specific detection of a disease in the early stage, with the added benefit of decreasing the chance of misdiagnosis.

4.2 Electronic Biosensors

In electronic biosensors, a biologically active layer is developed on the transducing element. Upon binding between the bioactive layer and the target molecule, in the case of ovarian cancer detection of CA-125, He4 or CEA, the electronic properties around the transducer undergo a measurable change that results in a read-out indicating the presence of the target molecule and at what concentration.

Electrical biosensors are the largest class of biosensors [12]. They utilise an electron transfer reaction under certain physical constraints that occur upon target binding. The transducer in these types of biosensors is generally an electrode or a set of electrodes made of a biologically compatible material, such as gold, fabricated on an insulating, rigid material substrate. Examples of primary electrical biosensing techniques are electrochemical impedance spectroscopy (EIS), cyclic voltammetry [13], and potential step chronoamperometry (CA) [14]. The application of these techniques to the development of biosensors for the detection of ovarian cancer will be discussed in the following section.

Before we proceed, a word must be said on the use of labels in biosensors. Many biosensors rely on the use of a label, an additional reporter molecule that is bound to the detected target biomarker, to either amplify or translate the signal generated by the transducer. Common choices for labels are enzymes, fluorescent molecules, magnetic beads and nanoparticles. Labels offer several advantages, including increased sensitivity and the ability to translate a generated electrical signal into a simplified read-out. Both of these characteristics are ideal in a point-of-care setting. Conversely, biosensors utilising labels increase the cost and create additional sample processing steps and handling requirements, coupled with an extended signal development time. They can also alter the binding capability and specificity of the biosensor. There is no clear right or wrong choice on the implementation of labels in biosensors, but it is a decision that must be weighed in the market development of any biosensor.

4.3 Ovarian Cancer Biosensors

Because CA-125 is such an important biomarker for ovarian cancer, there has been considerable work put into the development of a biosensor for its detection. We have developed biosensors for the detection of ovarian cancer biomarkers, specifically CA-125, due to its well-accepted and well-understood role, including its limitations, in ovarian cancer diagnosis and treatment [15]. We also developed an EIS biosensor to detect CA-125 using functionalised gold, nanometre-scale interdigitated electrode arrays that utilised an antibody-based detection layer to detect CA-125. With this sensor, we were able to detect CA-125 within a clinically relevant range.

Other electrical sensors have been developed to detect CA-125. Again, functionalised gold electrodes and a detecting monoclonal antibody to CA-125 were used to detect concentrations of CA-125 in spiked buffer, spiked serum and whole blood [16]. Cyclic voltammetry and differential pulse voltammetry were used to obtain the data. This sensor has several advantages over the EIS sensor developed by us. It is capable of detecting a higher and lower concentration range, and it is capable of detecting CA-125 in whole blood. However, the detection time for this sensor is 40 minutes, four times longer than the 10 minutes required by the EIS sensor, and while it is capable of detecting CA-125 in a number of different types of samples, none of the concentrations reported to be detected are within the key diagnostic range.

One of the first efforts to detect CA-125 was a biosensor that used cyclic voltammetry to monitor the changes in current over a small voltage range [17]. This sensor employed a sandwich immunoassay to detect CA-125 that included magnetic nanoparticles conjugated with a horseradish peroxidase (HRP)–anti-CA-125 complex as the sensing probes and thionine-HRP-doped silica nanoparticles as recognition elements. These particles contained both the enzyme and electron mediator labels, making the addition of labels into the sensor very convenient. Detection of CA-125 concentrations was accomplished after two 18-minute incubation steps.

As it is a newly discovered ovarian cancer biomarker, there has not been quite as much work done on the development of He4 biosensors. There are two biosensors that have been developed that we are aware of, and both utilise nanoparticles to develop the sensor read-out. There was a talk given at the Trends in Nanotechnology Conference in 2008 about a sensor utilising optically encoded nanoparticles to detect He4 and mesothelin in a high-throughput biosensor platform. Beads coupled with single-chain variable fragments specific to He4 and mesothelin are used to capture the He4 target protein. A fluorescent antibody is then used to generate an optical signal. Unfortunately, the range of concentrations detected is far above those required to determine the presence of disease, making it ineffective in a clinical situation. This work also appears to never have been published.

The second biosensor that we have come across in the literature is very promising [18]. It uses silver nanoparticles that have been conjugated with the He4 antibody as the sensing layer on a chip. Spiked solutions containing the He4 antigen, as well as clinical samples from patients, were then bound to the chip and local surface plasmon resonance (SPR) was used to develop the read-out. This sensor is robust and performs comparably to the ELISA for He4 detection. The read-out time is just 40 minutes.

The best chance for an electrical ovarian cancer biosensor to succeed in a clinical setting, detecting the disease in the early stages when the prognosis is still good, would be if the biosensor were capable of multiplexed detection of the disease's biomarker signature. We developed a multiplexed EIS sensor to detect CA-125, He4 and CEA, which is described in greater detail later in this chapter. This sensor utilised interdigitated electrode arrays with specific arrays designated with antibodies for each target antigen, along with an array that served as a control in order to account for nonspecific binding [19].

A recently reported sensor used magnetic nanoparticles coated with CEA and CA-125 antibodies to generate a voltage shift under manipulation by an external magnetic field [20]. This sensor contained five electrodes on a microfluidic device, and all of these were functionalised for the detection of a particular antigen. The target antigens, besides CEA and CA-125, were alpha-fetoprotein and CA 15-3. There was less than a 10% change in signal at the electrodes when they were exposed to proteins they were not functionalised to detect, creating great potential for specific detection of these target antigens simultaneously. This sensor is also capable of rapid detection ranging between 30 seconds and 5 minutes.

4.4 Development of a Practical Ovarian Cancer Biosensor

Our biosensor research originated from our desire to understand the interface between electronics and biomolecules. This work has led to the development of electronic sensors for several small molecules and proteins, including ovarian cancer biomarkers. We began a project for ovarian cancer detection by identifying two concurrent goals for the sensor: detection of multiple biomarkers for ovarian cancer within the clinically relevant range and development of an electronic platform capable of detecting low concentrations of these biomarkers.

We started the process of biosensor development by optimising our electronic sensing platform. This effort included overall design of a Secure Digital (SD) card format so that the chip would fit easily into a USB-powered reader. Our basic platform was contained on a plasma-cleaned 650 μm thick silicon chip coated with 1 μm of silicon dioxide. A thin layer of gold was then deposited on the chip, and standard photolithography techniques and deep reactive ion etch (DRIE) were used to fabricate the electrodes. Each chip contained eight gold electrodes, and each of these electrodes was comprised of 39 interdigitated fingers measuring 200 nm high, 1500 nm wide, and 800 nm long. Each finger was separated from a neighbouring one by a distance of 800 nm. Gold was chosen as the electrode material due to its high level of biocompatibility and because it provides an approachable surface for the chemical modifications necessary to develop a sensing layer on the electrode surface. Leading away from these eight sets of nanoscale interdigitated electrodes were eight contact pads and a reference electrode (Figure 4.1).

We chose to use nanometre-scale sensing elements for our work for several reasons. Sensitivity is of utmost importance in detecting disease states where

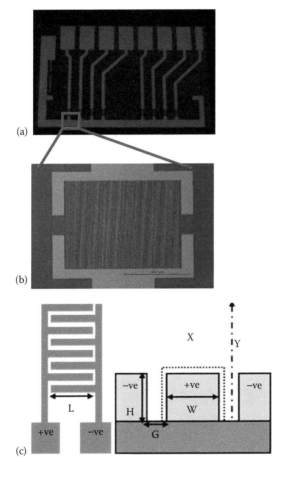

FIGURE 4.1
Overview of the biosensor chip. (a) Eight sets of gold electrodes are fabricated on a silicon chip coated with a thin layer of silicon dioxide. (b) Close-up view of an interdigitated electrode. (c) Schematic showing the critical dimensions: width (W), length (L), height (H) and gap (G) of an interdigitated electrode.

just an order of magnitude or a factor of 10 in concentration of a particular biomarker may separate a normal level from a diseased one. One way of enhancing the sensitivity of an electrical biosensor is to strengthen and concentrate the electric field. The use of nanoscale electrodes does just that. We used numerous computational models of the electric field and current density while adjusting parameters of our electronic platform and determined that the most important factor in enhancing the electric field, and thus increasing the sensitivity, is to decrease the distance between the electrode interdigitations. A related and important aspect of our electrode design was to consider surface area. Increasing the number of probe molecules attached to the transducing elements can also enhance sensitivity. With our interdigitated nanoelectrode design, we created a large surface area for the attachment of such molecules. Another reason to use nanoelectrodes is scalability. The silicon chip in our biosensor is designed for use for a single diagnostic in a point-of-care setting. This chip is fabricated using the conventional processes employed by the semiconductor industry and is estimated to cost less than a dollar.

The first step in developing the sensing layer was to create a self-assembled monolayer (SAM) on the gold electrodes that we then chemically activated. The antibody to the ovarian cancer biomarker of interest (CA-125, He4 or CEA) was then immobilised onto the SAM using a chemical linker (Figure 4.2). The details of these processes have been reported in our publications [19]. An important step in the functionalisation process is to prevent any nonspecific binding that may occur between the sensing layer and the substance being analysed. As such, any possible binding sites that remain must be blocked. A common choice for this in our work was the use of a nonprotein polymer,

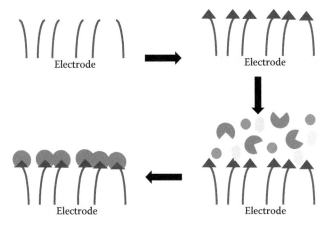

FIGURE 4.2
Overview of the biofunctionalisation scheme. A SAM is developed on the gold electrodes. An antibody is chemically attached to the SAM. A solution containing multiple antigens is incubated on the electrodes. Unbound antigens are washed away and the bound target antigen remains.

polyvinylpyrrolidone (PVP) that is biocompatible, water soluble, and forms thin layers.

After the electrodes were functionalised, a polydimethylsiloxane (PDMS) gasket was sandwiched between the electrodes and a Plexiglas channel that contained inlet and outlet feeds. This allowed us to create a channel that only passed over the electrodes, restricting the fluids to the electrode surface where the biosensing layer was located.

As mentioned earlier in the chapter, there are many detection spectroscopies that can be utilised for the detection of target molecules. For our electronic biosensors, we chose to utilise EIS to sense the binding of ovarian cancer biomarkers to the functionalised electrodes. EIS is an effective tool for investigating the electrochemical characteristics of modified transducers [21,22]. EIS has proven to be a sensitive method that can detect even a minute change in the capacitance and resistance of the transducer–detection layer interface [23,24]. The impedance measurement is based on measuring the nonlinear response of the circuit elements to the changing input alternating current (AC) frequency. The impedance is measured with an impedance analyser using the frequency response of the device. It can be expressed as the ratio of the system voltage phasor to the current phasor. Using phasors allows for incorporation of the phase angle between the current and voltage at different frequencies due to the nonlinear response.

To obtain EIS data, a small AC voltage of up to several hundred millivolts and a frequency range from tens of hertz to several megahertz were applied to the system and the response was collected. This broad sweep of frequencies allows the biosensor to detect fast events, such as electron transfer, and slow events, like diffusion-limited processes. The impedance measurement data contain a real and an imaginary component. These impedance components can be plotted as the imaginary versus the real to obtain a Nyquist plot (Figure 4.3). A Randles equivalent circuit can be fitted to the impedance data as a function of the frequency obtained through EIS measurement, as shown in Figure 4.3 [21,22,25,26]. There are other circuit models that can be fitted to EIS data, but the Randles circuit is the most common choice. Here, R_s is the resistance of the bulk solution, Z_W is the Warburg impedance that comes about as a result of the diffusion of ions in the solution between the electrodes, C_{DL} is the double-layer capacitance and R_{ET} is the electron transfer resistance.

R_{SOL} and Z_W are affected solely by the bulk solution. As such, these values will not vary with electrical activity occurring at the solution–electrode interface. C_{DL} and R_{ET}, on the other hand, are dependent on the dielectric and the electron transfer resistance at the electrode interface, respectively. These are the values that are altered when a binding event occurs between the functionalised electrode surface and the target biomolecule. It is this ability to differentiate between impedance changes in the bulk solution and impedance changes at the electrode interface that makes EIS an ideal, and the most common, electrochemical spectroscopic detection technique for biosensors.

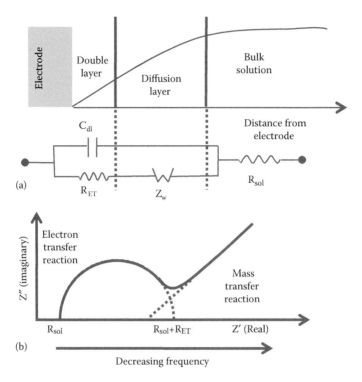

FIGURE 4.3

Electrochemical impedance spectroscopy. (a) Randles equivalent circuit used to model the interactions at the electrode surface, where the solution resistance is R_{SOL}, Z_W is the Warburg impedance, C_{DL} is the double-layer capacitance and R_{ET} is the charge transfer resistance. (b) Nyquist plot used to extract Randles circuit parameters.

Additionally, it is less damaging to biomolecules because it is measured over a small, narrow range of potentials.

It is possible to introduce a redox probe during EIS measurements. This is called faradaic electrochemical impedance spectroscopy (fEIS). In fEIS, a redox couple is used to probe the changes on the surface of the electrodes when a redox reaction occurs in the presence of a direct current (DC) bias that is either at or above the redox potential of the couple applied to the electrodes.

Our ultimate goal was to develop a biosensor capable of detecting multiple ovarian cancer biomarkers simultaneously. The first step towards this goal was to develop individual sensors capable of detecting each of the biomarkers: CA-125, He4 and CEA. We needed to ascertain that the biomarkers could be detected in the clinically relevant range. We used immunological affinity based on antigen–antibody binding. Monoclonal antibodies for each target were attached to the functionalised electrodes. Buffer solutions containing different concentrations of the complementary antigen were flowed through the channel located on top of the electrodes, and each given several minutes to bind. Excess antigen-spiked buffer was then washed away. A second buffer containing a detector antibody, complementary to the now bound antigen, was then introduced into the channel and given sufficient time to bind to the antigen–antibody complex. This additional bound antibody served to increase the specificity of the signal generated, as it would only bind to already bound antigen and it enhanced the sensitivity of the signal that arises from molecules bound to the biofunctionalised electrodes. In this process of successive attachment to the biofunctionalised layer, we developed a sandwich assay consisting of a stack of capture antibody, antigen and detector antibody on the electrodes on the chip, with each successive layer binding to the layer below it (Figure 4.4).

EIS data were collected after the incubation of each piece of the sandwich assay. Nyquist plots were produced and Randles equivalent circuits fitted to the plots in order to obtain values for the circuit components. The diameter of the semicircular portion of the Nyquist plot increases with the increasing number of bound molecules, enhancing the signal. A comparison of the R_{ET} values for each additional component of the sandwich assay shows an increase with each bound layer (Figure 4.5).

Let us consider the example of our biosensor for He4 detection [19]. It was capable of detecting He4 antigen in spiked buffer at concentrations between 1.5 and 25 ng/ml in serial twofold dilutions. This is well within the clinically relevant range believed to be required for diagnosis, ~5 ng/ml. The R_{ET} values obtained from the Nyquist plots upon He4 antigen binding are linear within

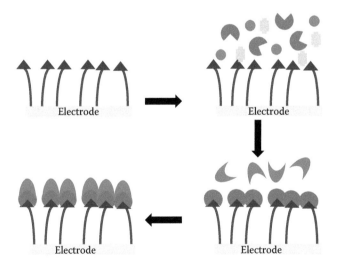

FIGURE 4.4
Sandwich assay amplification. Once the electrodes are biofunctionalised with a capture antibody, they are exposed to the target antigen. Once the target antigen is bound, a secondary detection antibody is incubated on the electrodes, which binds to the already bound antigen, simultaneously increasing the signal generated, by the sensor and the specificity.

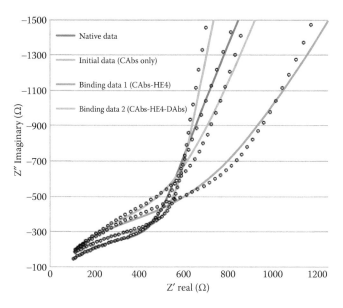

FIGURE 4.5

Nyquist plots obtained from a He4 sensor. The increase in diameter in the semicircular portion indicates an increase in the number of bound biomolecules. (From Whited, A. M., et al. An electronic sensor for detection of early-stage biomarker/s for ovarian cancer. *BioNanoScience,* 2012. 2(4): 161–170.)

this range, and the R_{ET} values from the additional binding of the detector antibody are directly proportional to the concentration of He4 bound to the electrodes. This clearly demonstrates that our biosensor is capable of specifically detecting He4 concentrations with the sensitivity required in a point-of-care situation. Similar sensors were developed by our lab for the detection of CA-125 and CEA, except that those did not rely on a secondary incubation of a detection antibody but rather detected the antigen by itself [15].

Having optimised the individual biosensors, we began the process of developing a biosensor platform capable of multiplexed detection of He4, CA-125 and CEA simultaneously. Our chip-based electrode system contained eight sets of nanoscale interdigitated electrodes. We divided the electrodes into four groups of two and functionalised each set of two electrodes with He4, CA-125 and CEA antibody (Figure 4.6). The final two electrodes were functionalised with ovalbumin and served as a negative binding control. A buffer solution containing all three target antigens at varying concentrations between 0.1 and 10 ng/ml, all within the clinically relevant range, was preincubated with the corresponding detection antibodies for each target antigen. Table 4.1 is a list of the exact composition of each sample. The concentration of the detection antibodies was kept at a constant 10 μg/ml in each sample. This solution, containing two parts of the sandwich assay, was passed

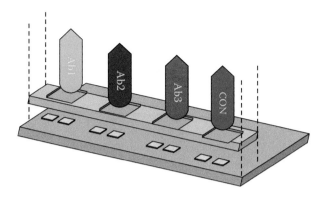

FIGURE 4.6
Multiplexed biosensor functionalisation scheme. Four groups of two electrodes each function-alised with capture antibodies for CA-125, CEA and He4, plus ovalbumin that served as a negative control. (From Whited, A. M., et al. An electronic sensor for detection of early-stage biomarker/s for ovarian cancer. *BioNanoScience*, 2012. 2(4): 161–170.)

TABLE 4.1

Composition of the Samples Used for Multiplexed Detection of Ovarian Cancer Biomarkers CA-125, CEA and He4

Sample	CA-125 (ng/ml)	CEA (ng/ml)	He4 (ng/ml)
S1	1	0.1	10
S2	0.1	10	1
S3	10	1	0.1

Source: From Whited, A. M., et al. An electronic sensor for detection of early-stage biomarker/s for ovarian cancer. *BioNanoScience*, 2012. 2(4): 161–170.

Note: Each sample was preincubated with monoclonal antibodies to all three biomarkers at a concentration of 10 µg/ml before the complexes were bound to the biofunctionalised electrodes.

through the channel over the electrodes and incubated for several minutes. The excess unbound antigen–detection antibody complex was then washed away and EIS data collected.

The difference in R_{ET} values obtained from the electrodes before and after incubation with the immunocomplex exhibited a unique pattern from each electrode set for each sample that was directly related to the elevated presence of each electrode's targeted biomarker (Figure 4.7). It is worthwhile to note that the electrodes biofunctionalised for the detection of CEA generated a saturated signal after a relatively low concentration of the target immuno-complex. Although we expect all of the electrodes to be saturated beyond a particular concentration of the target biomarker when all of the available

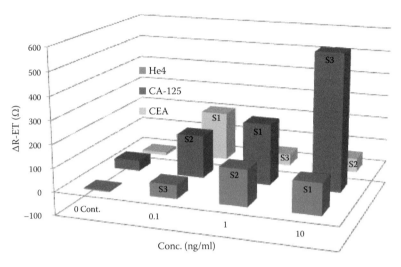

FIGURE 4.7

Multiplexed detection of ovarian cancer biomarkers. Each tested sample contained a mixture of CA-125, CEA and He4 of varying concentrations. Results show that each sample produced a unique signal from the electrodes while generating little to no signal from the control electrodes. (From Whited, A. M., et al. An electronic sensor for detection of early-stage biomarker/s for ovarian cancer. *BioNanoScience*, 2012. 2(4): 161–170.)

binding sites are occupied, this does seem to happen prematurely for these electrodes. The R_{ET} values obtained from the negative binding control electrodes functionalised with ovalbumin were significantly smaller for each sample than the electrodes specified for an ovarian cancer biomarker.

We tested these multiple multiplexed chips several times, and although they exhibited the same pattern of detection, the intrachip variability in signal was as high as 25%. However, the interchip signal variability was between 2% and 20%. From both our computational models of our biosensor platform and our experimental results, we know that the electric field and current densities are not the same across all electrode sets within the chip, with the largest differences coming from the electrodes on the edges of the chip. This may be responsible for the variation in the detected signals.

4.5 Future Directions in the Development of Ovarian Cancer Biosensors

As we have seen, ovarian cancer is a challenging disease to diagnose and treat. While the development of a biosensor capable of detecting the disease's biological thumbprint is still in the early stages, it has the promise of contributing to a decreased mortality rate from the disease and saving lives of countless women. Though much progress has been made to develop biosensors to replace the conventional assays used to detect ovarian cancer, much work still remains to be done. By and large, the ovarian cancer biosensors developed up to this point suffer from several deficiencies. The biggest problem and most central hurdle to their implementation in a clinical setting is that while they are able to detect the protein biomarker of choice in spiked buffer samples with accuracy, their detection specificity and sensitivity lag behind in detecting biomarkers in clinically relevant samples such as whole blood or whole serum. Second, they typically rely on a label-driven signal amplification step, increasing the time and cost of detection, both of which could be prohibitive to their implementation. Finally, they are not developed for a mobile platform. They rely on temperature-sensitive specialised reagents or chemicals that may not easily lend themselves to portable point-of-care situations, particularly in remote locations.

The limitations of the current ovarian cancer biosensors extend beyond the physical constraints. While this chapter has focused primarily on the detection of three of the well-known proteomic signatures of ovarian cancer, there are other viable targets for ovarian cancer biosensors. These could include ovarian cancer cells or other known protein biomarkers. There is plenty of room for improvement in our understanding of the disease and the best mechanism to detect it in the early stages, before it has metastasised beyond the primary site.

In addition to disease detection, biosensors may also play an important role in the treatment of ovarian cancer. The chemotherapy course for treatment of ovarian cancer is physically demanding and, like all chemotherapies, leaves the patient in a state of compromised immunity. The current course of action during this treatment period dictates that the patient travel to a brick-and-mortar laboratory to have blood drawn so that the level of CA-125 present in her blood serum may be determined in an effort to validate the effectiveness of the treatment. We imagine a world where a USB-based biosensor for detecting CA-125 in whole blood may be plugged into a patient's personal computer. From there, she pricks her finger, akin to the process diabetics use for a blood glucose sensor, and places a couple of drops onto the biosensor; her CA-125 level is measured, and the results immediately sent to her doctor.

Biosensors in general will play an important role in our evolving health ecosystem with their ease of use and reduced cost and time compared with traditional laboratory testing. With our increasing understanding of genetics and the need for personalised medicine, biosensors are poised to have a large impact on the way we view disease diagnosis. While many disease states will benefit from this surge towards rapid, cheap disease diagnosis, we are truly hopeful that the work that has been done on ovarian cancer detection will ultimately lead to a reduced mortality rate for this ruthless disease.

References

1. Yancik, R. Ovarian cancer: Age contrasts in incidence, histology, disease stage at diagnosis, and mortality. *Cancer*, 1993. 71(S2): 517–523.
2. Goff, B. A., et al. Ovarian carcinoma diagnosis. *Cancer*, 2000. 89(10): 2068–2075.
3. Ries, L. A. G., et al. SEER cancer statistics review, 1975–2005. Bethesda, MD: National Cancer Institute, 2008, pp. 1975–2005.
4. Hellström, I., et al. The HE4 (WFDC2) protein is a biomarker for ovarian carcinoma. *Cancer Research*, 2003. 63(13): 3695–3700.
5. Andersen, M. R., et al. Use of a symptom index, CA125, and HE4 to predict ovarian cancer. *Gynecologic Oncology*, 2010. 116(3): 378–383.
6. Bast Jr., R. C., et al. Reactivity of a monoclonal antibody with human ovarian carcinoma. *Journal of Clinical Investigation*, 1981. 68(5): 1331.
7. Yin, B. W., and K. O. Lloyd. Molecular cloning of the CA125 ovarian cancer antigen: identification as a new mucin, MUC16. *Journal of Biological Chemistry*, 2001. 276(29): 27371–27375.
8. Bingle, L., et al. The putative ovarian tumour marker gene HE4 (WFDC2), is expressed in normal tissues and undergoes complex alternative splicing to yield multiple protein isoforms. *Oncogene*, 2002. 21(17): 2768–2773.
9. Ranganathan, S., et al. The whey acidic protein family: A new signature motif and three-dimensional structure by comparative modeling. *Journal of Molecular Graphics and Modelling*, 1999. 17(2): 106–113.
10. Drapkin, R., et al. Human epididymis protein 4 (HE4) is a secreted glycoprotein that is overexpressed by serous and endometrioid ovarian carcinomas. *Cancer Research*, 2005. 65(6): 2162–2169.
11. Gold, P., and S. O. Freedman. Specific carcinoembryonic antigens of the human digestive system. *Journal of Experimental Medicine*, 1965. 122(3): 467–481.
12. Manz, A., et al. *Bioanalytical Chemistry*. Vol. 69. Singapore: World Scientific, 2004.
13. Ianniello, R. M., et al. Differential pulse voltammetric study of direct electron transfer in glucose oxidase chemically modified graphite electrodes. *Analytical Chemistry*, 1982. 54(7): 1098–1101.
14. Bartlett, P. N., et al. Kinetic aspects of the use of modified electrodes and mediators in bioelectrochemistry. *Progress in Reaction Kinetics*, 1991. 16(2): 55–155.
15. Whited, A. M., et al. Label-free electrochemical impedance detection of ovarian cancer markers CA-125 and CEA. In *MRS Proceedings*. Cambridge: Cambridge University Press, 2009.
16. Das, J., and S. O. Kelley. Protein detection using arrayed microsensor chips: Tuning sensor footprint to achieve ultrasensitive readout of CA-125 in serum and whole blood. *Analytical Chemistry*, 2011. 83(4): 1167–1172.
17. Tang, D., et al. Nanoparticle-based sandwich electrochemical immunoassay for carbohydrate antigen 125 with signal enhancement using enzyme-coated nanometer-sized enzyme-doped silica beads. *Analytical Chemistry*, 2010. 82(4): 1527–1534.
18. Yuan, J., et al. Detection of serum human epididymis secretory protein 4 in patients with ovarian cancer using a label-free biosensor based on localized surface plasmon resonance. *International Journal of Nanomedicine*, 2012. 7: 2921.

19. Whited, A. M., et al. An electronic sensor for detection of early-stage biomarker/s for ovarian cancer. *BioNanoScience*, 2012. 2(4): 161–170.

20. Tang, D., et al. Magnetic control of an electrochemical microfluidic device with an arrayed immunosensor for simultaneous multiple immunoassays. *Clinical Chemistry*, 2007. 53(7): 1323–1329.

21. Bard, A. J., and L. R. Faulkner. *Electrochemical Methods: Fundamentals and Applications*. Vol. 2. New York: Wiley, 1980.

22. Stoynov, Z., et al. *Elektrokhimicheskii impedans*. Moscow: Nauka, 1991.

23. Athey, D., et al. A study of enzyme-catalyzed product deposition on planar gold electrodes using electrical impedance measurement. *Electroanalysis*, 1995. 7(3): 270–273.

24. Nahir, T. M., and E. F. Bowden. The distribution of standard rate constants for electron transfer between thiol-modified gold electrodes and adsorbed cytochrome c. *Journal of Electroanalytical Chemistry*, 1996. 410(1): 9–13.

25. Ershler, B. Investigation of electrode reactions by the method of charging-curves and with the aid of alternating currents. *Discussions of the Faraday Society*, 1947. 1: 269–277.

26. Randles, J. E. B. Kinetics of rapid electrode reactions. *Discussions of the Faraday Society*, 1947. 1: 11–19.

5

Sensors and Sensor Systems for Harsh Environment Applications

Andrea De Luca, Florin Udrea, Guoli Li, Yun Zeng, Nicolas André,
Guillaume Pollissard-Quatremère, Laurent A. Francis, Denis Flandre,
Zoltan Racz, Julian W. Gardner, Syed Zeeshan Ali, Octavian Buiu,
Bogdan C. Serban, Cornel Cobianu and Tracy Wotherspoon

CONTENTS

ABSTRACT This chapter focuses on the development of sensors and sensor systems for harsh environments. Among various applications with high commercial impact, combustion optimisation and emission control in small-scale boilers is considered as an illustrative application which can benefit from the employment of a multimeasurand sensor system able to cope with harsh environment conditions. More specifically, silicon on insulator (SOI) is proposed as common technology platform for the realisation of a diode temperature sensor, a thermal flow sensor, a capacitive humidity sensor, a chemiresistive oxygen sensor, a diode ultraviolet (UV) photosensor, and a non-dispersive-infra-red (NDIR) carbon dioxide sensor. Considerations regarding circuitry, packaging and system integration and testing are also included, along with a detailed analysis of each single sensing device.

5.1 Introduction

Research and development of sensors able to operate in harsh environments is strongly driven by market needs resulting from a growingly stringency in environmental regulations and people safety legislation. One might simplistically associate the term *harsh* exclusively with high ambient temperature, but it is actually possible to identify four physical domains (thermal, mechanical, chemical and electromagnetic) that can contribute to the harshness of an environment. Furthermore, one has to bear in mind that different physical domains can interact with each other with complex cause–effect relationships, and that each single physical domain must be considered not only in steady state or slowly varying conditions, but also, and probably with even more concern, in a transient perspective. Under these conditions, depending on the specific application, a variety of measurands in different physical domains might be of interest (for direct monitoring and/or compensation purposes), such as temperature, flow, pressure, acceleration, radiation, gases (CO, CO_2, NO_x, H_2S, O_2, O_3, water vapour, etc.) and particles.

5.1.1 Harsh Environment Applications

There is an abundance of applications (combustion optimisation and emission control in automotive, marine and aerospace, as well as in industrial and small-scale boilers; carbon capture and sequestration; oil and gas storage and sequestration; smart infrastructures; space exploration; etc.) that would benefit from the employment of sensors able to cope with harsh environment conditions. Other emerging economic areas would also flourish if such sensors became available with an attractive price/performance ratio.

Hereafter, combustion optimisation in small-scale boilers is presented as an illustrative application for a multimeasurand sensor system able to cope with harsh environments. Currently, the domestic boilers market (5 million domestic boilers sold per year in the European Union, with an annual growth rate around 15%) is divided into two categories: (1) highly efficient (HE) premix condensing systems and (2) standard efficient (SE) atmospheric systems. The HE boilers are expected to prevail over the SE systems in the near future. By being based on the premix system (1:1 gas/air ratio control), they ensure safe handling (if there is no air, then no gas will be pulled out from the system), fast dynamic response, low cost and modulation with premix burner [1]. They additionally guarantee lower gas emissions and ultimately higher efficiency, in comparison with SE boilers, due to their autoadaptability feature. This is achieved by inserting gas sensors above the burner to measure or detect the combustion quality, and by trimming the gas flow through a motor-driven throttle. The gas sensor can be either an O_2 or a CO_2 sensor or a multimeasurand structure capable of detecting both gases.

The O_2 and CO_2 sensors need to cope with operating temperatures up to 225°C and water vapour presence in the gas composition above 10% in volume. HE boilers, taking benefit of the flue condensation, also have the option for gas measurement after water vapour condensation. A reduction of the operative temperature is thus possible, but the relative humidity (RH) can be as high as 100%. Therefore, a need for simultaneous monitoring of temperature and RH, together with either O_2 and CO_2 (ideally, both) concentration, emerges. For maximum enhancement of the combustion process efficiency, total gas flow also has to be measured. The system might also be complemented with the presence of a UV photodetector for optical flame safety monitoring purposes.

Furthermore, low-power consumption, reliability and extended lifetime (>3 years) are also requirements linked to overall system efficiency. To date, HE boilers include only a CO_2 sensor to achieve their autoadaptability feature because no multimeasurand solution able to cope with the aforementioned harsh conditions is available off the shelf. Concurrent gas flow, RH, temperature, UV, O_2 and CO_2 detection is obviously a significant benefit provided that the extra costs resulting from the implementation of such a system do not detrimentally affect the commercial appeal of the product.

5.1.2 Sensing Technologies for Harsh Environments

Now, the issue is understanding which technologies are capable of satisfying this market demand. Bulk silicon microsensors are a reality. Price-wise and performance-wise, very competitive products are easily available off the shelf. However, if operating at temperatures in excess of 125°C, in corrosive atmospheres, in the presence of radiation or mechanical stresses, they can incur serious reliability problems. Silicon-on-insulator (SOI) devices, although not too different from bulk silicon–based sensors from a chemical and mechanical perspective, are provided with an extended working temperature range (up to 300°C) and radiation hardness. In the last decade, both industry and academia have put a lot of effort in the development of new materials able to overcome these limitations. Silicon carbide (SiC), gallium nitride (GaN) and diamond are certainly extremely attractive in terms of intrinsic properties (Table 5.1). Devices based on these technologies can be operated at temperatures in excess of 300°C and are chemically and mechanically more robust than silicon. However, low availability, processing costs and yield severely hinder their mass market use, relegating them to niche and high-end applications, not accessible by bulk Si or SOI (e.g. $T > 300$°C). Furthermore, circuitry integration is possible and easily achievable in silicon technologies, while it is much more challenging for SiC, GaN and diamond. In this scenario, SOI is clearly an excellent compromise.

SOI technology offers significant advantages in four areas: sensors, harsh environment electronics, high-power electronics and high-speed electronics. The employment of SOI substrates is also particularly attractive for

TABLE 5.1

Material Properties

	Bulk Si	SOI	SiC[a]	GaN	Diamond
Bandgap (eV)	1.1 (I)	1.1 (I)	2.4–3.2 (I)	3.47 (D)	5.5 (I)
Thermal expansion (10^{-6}/K)	2.6	2.6	2.9–3.3	5.6	1
Thermal conductivity (W/cm K)	1.45	1.45	3.6–4.9	1.3	20
Melting point (K)	1,690	1,690	3,100	2,400	4,200
Young's modulus (GPa)	170	170	350–700	220–330	1,000
Hardness (kg/mm²)	1,000	1,000	3,300	1,450	10,000
Chemical resistance	Poor	Poor	Excellent	Good	Good, but burns
Commercial availability	Excellent	Excellent	Medium	Poor	Very poor
Wafer size	12 in.	8 in.	6 in.	6 in.	2 in.

Source: Kroetz, G. H., et al., *Sensors and Actuators A* 74: 182–189, 1999; Cimalla, V., et al., *Journal of Physics D* 40: 6386–6434, 2007.

Note: I, indirect; D, direct.

[a] Properties vary between 3C-SiC, 4H-SiC and 6H-SiC.

the fabrication of suspended structures in microelectromechanical system (MEMS) sensing devices, since the buried oxide acts as an etch stop for frontside and backside etching techniques. The buried oxide is also an excellent isolation layer (both thermally and electrically) between the thin silicon and the silicon substrate. Furthermore, structures realised within the thin single-crystal silicon are characterised by superior mechanical and electrical properties if compared with those realised in polysilicon. The arguments in favour of SOI for electronic applications are also related to the buried oxide/thin silicon sandwich structure. Transistors are vertically isolated from the substrate and horizontally isolated from each other by the buried oxide and the deposited oxide, respectively. This electrical isolation, coupled with the small thickness of the active silicon layer, results in (1) reduced parasitic components – as a consequence, SOI circuits can be faster (for the same supply voltage and circuit topology) than those realised in bulk complementary metal oxide semiconductor (CMOS) or can be less 'power hungry' (for the same speed and circuit topology); (2) enhanced latch-up immunity; (3) very small leakage currents – as a consequence, SOI circuits can operate up to 300°C; (4) radiation hardness – the buried oxide acts as a barrier, strongly reducing the radiation-induced photocurrents; and (5) superior scalability opportunities – miniaturisation of both transistor and circuits is easier than in bulk silicon technology.

5.2 SOI Multisensor Technology Platform

In order to implement within the same technology a multimeasurand sensing system for harsh environment applications, the choice of the transduction principles to exploit for measuring each single quantity of interest is critical. Simplicity and robustness are the main criteria to consider. Thermal-based transduction principles satisfy these requirements, and can be implemented with the structure depicted in Figure 5.1. Part of the bulk silicon of the SOI substrate is etched away through deep reactive ion etching (DRIE) to form ~5 µm thick full membranes (more mechanically robust than etched-through structures, such as cantilever or beam-supported membranes). DRIE guarantees nearly vertical cavity side walls, and thus aggressive device miniaturisation, while exploiting the buried oxide as effective etch stop. Tungsten high-temperature metalisations (more electromigration resistant than aluminium or polysilicon) are employed to realise the interconnects and the microheaters, fully embedded within such membranes for enhanced thermal isolation. Underneath the heating elements, the thin single-crystal silicon layer can be used to fabricate a temperature sensor (e.g. a diode or a thermopile) to accurately monitor the microheater temperature. Additional metal layers above the heating elements can be also introduced to realise heat-spreading plates, with the purpose of improving the uniformity of the temperature distribution in the 'hot zone' of the device. Above the passivation layer, gold electrodes (more chemically inert than aluminium) can be also provided in order to allow electrical characterisation of a gas-sensing layer deposited on top of them. Off membrane, not affected by the high temperature reached by the heating elements due to the thermal isolation offered by the dielectric membrane, a secondary diode temperature sensor

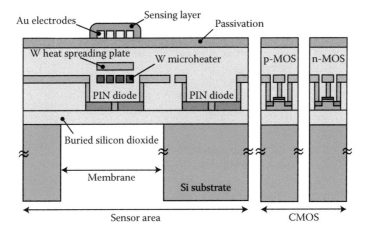

FIGURE 5.1
Schematic depiction (not to scale) of SOI CMOS MEMS technology platform.

can be placed to monitor ambient temperature variations, and provide the whole multisensor structure with ambient temperature compensation capabilities. Additional electronics, for active element driving or read-out, can be also implemented to realise a smart multimeasurand sensing system.

Hereafter, temperature, flow, RH, O_2, UV and CO_2 sensors and their implementation within the described SOI CMOS MEMS technology is discussed in detail and their operations in harsh conditions are presented.

5.2.1 Temperature Sensor

The presence of temperature sensors in a thermal-based multisensor structure aims to fulfil a dual need: (1) ambient temperature monitoring and compensation – by placing the sensor off membrane, and (2) heating element temperature monitoring – by placing the sensor on membrane. Given that the heating elements might need to operate hundreds of degrees above ambient temperature, the temperature sensors have to withstand temperatures well in excess of 300°C (beyond the maximum junction temperature for SOI devices). Diodes are the perfect candidates for temperature measurements in such an application: (1) they can be extremely small, (2) they do not offer a thermal bridge between hot and cold zones of the chip (conduction losses are thus minimised in comparison with thermopiles), (3) they are quite linear, (4) they have a wide range and (5) their sensitivity can be significantly enhanced by putting them in array form. Detailed analysis of SOI diode temperature sensors can be found in [2–4]. The voltage drop across an ideal p-n junction is given by

$$V = \frac{kT}{q} \ln\left(\frac{I_d}{I_s(T)} + 1 \right) \tag{5.1}$$

where I_s is the saturation current, k is Boltzmann's constant, T is the temperature, q is the electron charge and I_d is the driving current. The saturation current can be expressed as

$$I_s(T) = CA_j T^\eta e^{-\frac{qV_g}{kT}} \tag{5.2}$$

Here, C is a constant (dependent on the density of states, effective mass and mobility of carriers, doping density and lifetime), A_j is the junction area, η is a process-dependent parameter and V_g is the extrapolated bandgap voltage at 0 K. Given the high quality of the employed SOI wafers (no epilayers are involved) and the small thickness of the silicon layer (~250 nm), the lifetime is very high and the volume of the depletion region is very small, and a very wide working temperature range (–200°C to 750°C) can be achieved, as shown in Figure 5.2.

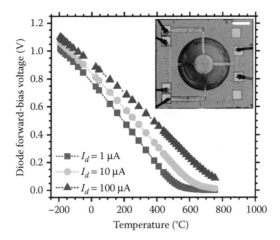

FIGURE 5.2
Measured diode forward-bias voltage vs. heater temperature of a p+/p– well/n+ diode with 5 μm² junction area for different driving currents (1, 10 and 100 μA). Inset: UV-filtered optical micrograph of a microhotplate chip, with diodes underneath the glowing microheater operated at high temperature (<750°C). Scale bar is 200 μm.

5.2.2 Flow Sensor

It has been long known how to design thermal flow sensors [5], and their development, in parallel with technology advancements, has seen significant progress in recent years, due to (1) the absence of moving parts, (2) relatively simple fabrication in comparison with other transduction principles and packaging schemes and (3) possible CMOS-compatible implementation. However, in literature there is a surprising lack of reports dealing with flow sensor operations in harsh environments. Chen et al. [6] deal with high RH levels, and De Luca et al. [7] present results at both high temperature and RH levels, as well as a novel multidirectional sensing arrangement summarised hereafter. An optical micrograph of the fabricated sensor is presented in Figure 5.3. The thermoelectric flow sensor chip comprises a 60 μm diameter microheater with a diode temperature sensor underneath. Eight thermopiles are symmetrically arranged in pairs around the microheater with both hot and cold junctions embedded within the membrane. Each thermopile is formed by nine single-crystal Si p+/n+ thermocouples. Tungsten is used as an interconnect between the thermoelements, in order to avoid the formation of junctions with rectifying behaviour. Effective thermal isolation and robust mechanical support for the heating and the temperature sensing elements are achieved by a circular (600 μm diameter) dielectric membrane. All pads were placed far away from the sensing area in order to avoid interaction between the airflow and the bonding wires. The chip is provided with a diode temperature sensor placed on the substrate for ambient temperature monitoring and compensation purposes (identical to the one underneath the

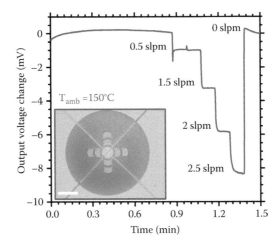

FIGURE 5.3

Temperature-compensated output voltage change of the flow sensor operated in thermoresistive mode at an ambient temperature of 150°C. Inset: Optical micrograph of the flow sensor membrane area. Scale bar is 100 µm.

microheater). The flow sensor was tested in harsh environment conditions by sealing a small flow channel (2 mm × 2 mm cross section) on top of the package and mounted within a custom-made high-temperature gas system. The sensor was operated in a thermoresistive constant current mode, meaning a constant value of current was injected in the heating element while monitoring the voltage drop across it. The flow sensor was first tested at 150°C in dry air in the volumetric flow rate range of 0–2.5 slpm (standard litres per minute). During the test, a linear drift of the sensor output, due to the slow thermal inertia of the system (package + flow channel within the gas system oven), was observed through the on-chip reference diode and was analytically removed in data postelaboration. A very promising response to flow rate is shown in Figure 5.3, proving the flow sensor to be suitable for high-temperature applications. Furthermore, the sensor was also characterised in the volumetric flow rate range of 0–0.7 slpm and humidity was varied in the range of 0%–75% in volume. A clear response [7] to different flow rates and a negligible humidity effect were observed.

5.2.3 Humidity Sensor

Miniaturised humidity sensors are largely studied. Table 5.2 shows a comparison between our work and typical humidity sensors, that is, porous Al_2O_3-based and polyimide-based sensors. These sensors use capacitive sensing between planar electrodes, whose response varies with water vapour absorption for polymer, or adsorption and capillary condensation for porous alumina.

TABLE 5.2

Humidity Sensor Performance Comparison

Reference	Best Sensitivity	Sensing Layer	Response Time	T Range (°C)	RH Range (%)
31	~70 fF/%RH/ mm²	Polyimide	15 s[a]	–40 to 120	0–100
32	—	Polymer	8 s[b]	–40 to 125	0–100
33	~20 fF/%RH/ mm²	Polymer	6 s[c]	–40 to 190	0–100
34	0.75 pF/%RH/ mm²	Polyimide	3 s	10–70	25–85
35	25 kHz/%RH/ mm²	Polyimide	—	25–80	25–95
11	15 pF/%RH/ mm²	Porous Al_2O_3	—	5–95	40–90
This work	2 pF/%RH/ mm²	Al_2O_3	0.5 s[d]	25–150	0–95

[a] 30%–90% RH.
[b] $\tau_{63\%}$ of a step function.
[c] $\tau_{63\%}$ 50%–0% RH.
[d] 45%–100% RH.

The humidity sensor presented here, which intends to withstand a higher-temperature regime, is based on aluminium oxide as well. This ceramic material has been demonstrated to be highly selective for moisture measurement [8,9] and an excellent material for measurement of moisture in most industrial gases [10], while polymer-based sensors cannot withstand high temperatures, as the Sensirion© SHT 21 specified only up to 125°C. Moreover, our sensor has a short time response, no longer limited by the diffusion of water inside of the material, but only on its surface as the Al_2O_3-sensitive layer is dense and thin.

The 25 nm thick hydrophilic Al_2O_3 layer is deposited during a unique post-process step by thermal atomic layer deposition (ALD) at 150°C from trimethylaluminium (TMA) and water vapour (H_2O) precursors. The deposition conditions, compatible with either full-wafer or packaged suspended membrane sensors, allow customisable functionalisation on different dies. Figure 5.4 presents the interdigitated electrodes (IDEs) functionalised by ALD to improve the humidity sensing. Moreover, as the humidity IDEs are located at the surface of the membrane, they allow the presence of a microheater and a thermodiode for temperature sensing, both embedded below the IDEs. Indeed, to unambiguously determine the atmospheric %RH point (the relative humidity (RH) level is commonly expressed in percentage, where 0% means totally dry air and 100% means saturated wet air), the local temperature knowledge is required, and the microheater can be used for temperature dependence cancellation, to reduce the recovery time after condensation or to decrease the drift due to the formation of chemisorbed OH groups [11] and contamination by dust.

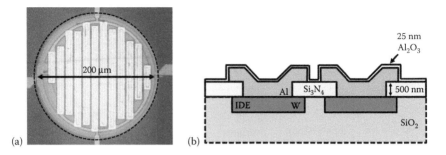

FIGURE 5.4
On-membrane IDEs with %RH-sensitive ALD coating: (a) top view and (b) cross section view.

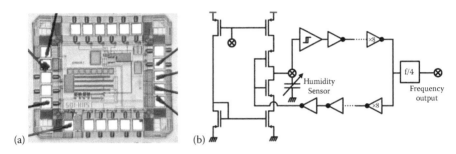

FIGURE 5.5
(a) SOI read-out circuitry for membrane operation. (b) Read-out schematic for %RH C-to-F transducer.

The read-out circuit, built in the same SOI CMOS technology as the membrane, is designed to operate between –55°C and +225°C (Figure 5.5a). The circuit provides drivers for the microheater and the diode temperature sensor. The circuit also includes a capacitive humidity sensor read-out (see schematic in Figure 5.5b) based on a ring oscillator that converts the capacitance value into frequency. The humidity variations are transduced into capacitance and then converted to oscillating voltage period variations, with a 200 μW power consumption. For high (low) %RH levels, the IDE capacitance increases (decreases) due to the amount of adsorbed water vapour, making the oscillating voltage period increase (decrease). At 25°C, the sensitivity to humidity is equal to ~2.5%/%RH for ALD-coated sensors and ~80 ppm/%RH for uncoated ones. The frequency output shows an accuracy level of ±2% RH [12]. This microsensing system, composed by the IDEs and read-out circuit was successfully tested up to 150°C.

5.2.4 Oxygen Sensor

During the last decades, several types of ceramic materials were employed as sensing layers in resistive O2–detecting structures aiming to work at high temperature for engine combustion control. TiO_2, Ce_2O_3, Ga_2O_3 and $SrTiO_3$ were among the n-type semiconducting metal oxides reported in literature [13] as showing good chemiresistive (changes in the resistance value of the sensing layer are correlated to changes in gas concentrations) oxygen response. Unfortunately, due to the severe variation with temperature of their conductivity, they are not a reliable choice for sensors working in high-temperature environments. However, when the $SrTiO_3$ was doped with iron atoms, the temperature coefficient of resistance (TCR) of the new $SrTi_{1-x}Fe_xO_{3-\delta}$ ($STFO_x$) changed from negative to positive, while increasing the Fe concentration [14]. For a certain atomic ratio of Ti to Fe, in the temperature range from 450°C up to 650°C, $STFO_x$ was shown to have TCR = 0; depending on the manufacturing method, $STFO_x$ exhibits TCR = 0 for either 35% Fe or 65% Ti (STFO35). A similar result was reported by Neri et al. for 60% Fe and 40% Ti (STFO60) [15].

Over the same period different synthesis procedures were tried for the preparation of the single (TiO_2) or composite (STO, STFO) metal oxide powders to be used as starting slurry materials for the deposition of thick and thin films aimed at O_2 sensing. As an example, we can mention electrospinning [16] and self-propagating high-temperature synthesis (SHS) [15]. Such methods require high-temperature annealing for the structural and compositional stabilisation of the final $STFO_x$ sensing layer, while the SHS method uses a self-propagating high-temperature sintering process, taking place at about 2300°C.

Recently, we have shown how to obtain $STFO_x$ sensing films starting from the $STFO_x$ powder prepared by the sonochemistry method [17]. Sonochemistry is a relatively new synthesis method able to generate a large family of organic, inorganic and biomaterials with controlled nanostructuring and large surface area, that can be used in multiple industrial and biomedical applications [18]. Within a sonochemical reactor, the electrical energy is converted by a piezoelectric generator into the vibration of a metal probe which is immersed in the aqueous liquid containing the precursors of the final reaction product. The transferred acoustic energy generates the transient acoustic cavitation in the reacting liquid evidenced by the appearance, development and finally collapse of the small bubbles. It has been already proven that the hot gas inside the bubble can have a temperature up to 5000 K and a pressure up to 1000 bars [18]. The energy released by the collapse of hot gas bubbles is the driving force of free radical generation, like hydrogen atoms (H•) and hydroxyl groups (OH•), which are at the origin of the enhanced chemical reactions taking place within the bubble and its vicinity, and are thus generating nanostructured powders. For the synthesis of the $STFO_x$ powder by the sonochemical method, the $Sr(NO_3)_2$ and $Fe(NO_3)_3$ precursors were initially diluted in water and then mixed with TiO_2 nanopowder for obtaining a homogeneous solution [19,20]. Then, this solution was slowly added to a solution of 4 M NaOH, under stirring. The final solution was exposed to a high-acoustic-intensity sonication process for about 1.5 h. The resulting powder was washed, filtered and dried in air, and finally annealed at 1000°C for 2 h. The spectroscopic evaluations have shown that the above sono-STFO$_x$ powder had a unique phase with the following stoichiometry: SrTi$_{0.6}$Fe$_{0.4}$O$_3$ (STFO40). For the preparation of the sono-STFO40 slurry – which will be used in the deposition of the O$_2$ sensing layer – the following formulation was performed: STFO40 (powder 60% w/w) + terpineol (solvent, 30% w/w) + ethyl cellulose (binder, 5% w/w) + capric acid–caprylic acid (equimolecular mixture surfactant, 5% w/w). Within the EU FP7 SOI-HITS project, the O$_2$ sensing results were obtained by using both sono-STFO40 and commercial STFO60 acquired from Neri et al. [21]. Both types of STFO$_x$ slurries were deposited by a dip-pen nanolithography (DPN) method [22] on the sensing area of the suspended SOI microhotplate membrane, as shown in the inset of Figure 5.6. By using a home-made experimental setup for functional evaluation of the SOI-CMOS-compatible resistive O$_2$ sensor (consisting of mass flow controllers, a bubbler, an ultrasmall testing chamber with controlled humidity and a computer), we obtained a similar O$_2$ response on both types of STFO$_x$, when oxygen was varied between 0% and 16% in the nitrogen ambient at RT [23]. The decrease of the STFO$_x$ resistance as a function of the O$_2$ concentration increase has proven the p-type semiconductor behaviour of the sensing layer, in accordance with previously reported results. The response time was less than 6 s, while the recovery time was less than 50 s, proving its efficiency for potential combustion control in different types of domestic and industrial boilers. For this application, the O$_2$ sensor encapsulated by Microsemi for complying with high-temperature operation is positioned in a boiler region near the combustion place, and where the temperature could not be much above 100°C. For this reason, by using a home-made experimental setup, we have evaluated the O$_2$ sensor response for ambient temperature in the range from 25°C to 160°C, while O$_2$

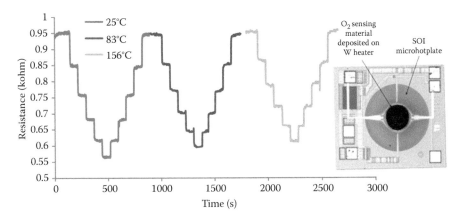

FIGURE 5.6
Response of the encapsulated SOI-CMOS-compatible resistive O$_2$ sensor as a function of O$_2$ concentration from 1% to 32% and the ambient temperature varying from 25°C to 156°C.

concentration was varied in discrete steps from 1% to 32% in nitrogen, as shown in Figure 5.6. These preliminary experimental results have shown an excellent p-type response, a small hysteresis and a stable enough behaviour with respect to ambient temperature in the entire investigated range. More work should be done for sensor stability and baseline drift control. This is the first time that a low-power (<80 mW) O_2-resistive sensor has been developed in a SOI-CMOS-compatible technology, proving that such a technology platform can be employed for the detection of a large family of gases present in harsh environments by integrating the appropriate sensing layer.

5.2.5 Photosensor

The SOI PIN diode can be used as an efficient photodetector. Controlling the depletion extension in the intrinsic (I) or low-doped intermediate region, defined by its intrinsic length (L_i), carriers generated by light radiation in the I region can be quickly separated by the lateral electric field and collected efficiently [24].

A cross section and top view of the SOI diodes suspended on the micro-hotplate platform with a bottom mirror are presented in Figure 5.7, where the diodes lying on the substrate are used as reference [25]. As the ~5 μm thick membrane is transparent to light wavelengths above 450 nm, optical reflection from the bottom mirror will occur in the multilayer stack to improve light absorption in the on-membrane thin-film PIN diode.

The normalised experimental photo- and dark currents for diodes with L_i = 5, 10 and 20 μm are plotted in Figure 5.8 as a function of temperature, T, from room temperature (RT) to 200°C, under a reverse anode bias, V_d, of –2.0 V and high-power 596 nm yellow light-emitting diode (LED) illumination.

The measured dark currents increase with temperature, because dominated by thermal volume (Shockley–Read–Hall) generation of current I_{rg} in the I region, as given in Equation 5.3 [24]. The currents are hence proportional to the depletion length, L_d, which is approximately equal to L_i in this case.

$$I_{rg} = \frac{q \cdot n_i}{2 \cdot \tau_{\text{eff}}} \cdot W \cdot t_{Si} \cdot L_d \cdot \left(e^{\frac{V_d}{2 \cdot U_t}} - 1 \right) \tag{5.3}$$

where τ_{eff} is the carrier's effective lifetime, n_i is the intrinsic carrier's concentration, W is the device width, t_{Si} is the thickness of active Si film and U_t is the thermal voltage. The main dependence on T is linked to n_i.

A quantitative difference of the experimental dark currents between the on-membrane and on-substrate diodes is induced by the carrier's effective lifetime, τ_{eff}, due to the post-CMOS process used to fabricate the circular membrane.

As shown, up to about 200°C, the photocurrents of the SOI PIN diodes can still be discriminated from the increasing dark currents, and the

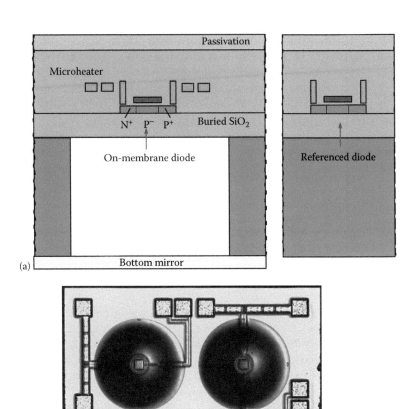

FIGURE 5.7
(a) Cross section and (b) top view of the SOI PIN diodes suspended on a microhotplate platform and the on-substrate reference diodes.

on-membrane photodiodes continuously achieve higher optical response compared with the reference ones on substrate.

Device responsivity, R, is defined in Equation 5.4 [26]:

$$R = \frac{I_{tot} - I_{dark}}{P_{in}} = \frac{I_{ph}}{P_{in}} \tag{5.4}$$

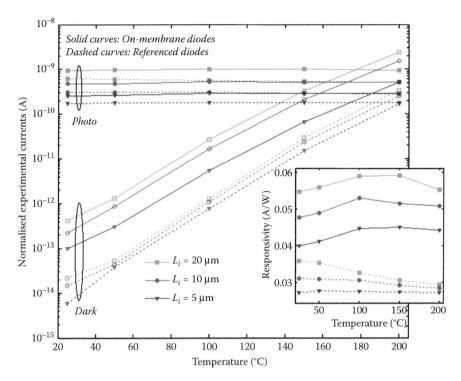

FIGURE 5.8
Normalised experimental photo- and dark currents of photodiodes (with $L_i = 5$, 10 and 20 μm) with increasing temperature up to 200°C, under –2.0 V reverse bias and 596 nm yellow LED illumination. Inset: Extracted device responsivities.

where I_{ph} is the photocurrent flowing through the I region and P_{in} is the optical power incident to the total device area.

At RT, the responsivities (given by the inset in Figure 5.8) of the on-membrane photodiodes with $L_i = 5$, 10 and 20 μm are 0.04, 0.048 and 0.055 A/W, respectively. R increases with L_i due to the increasing percentage of the photosensitive area (defined by L_i) over the total device area (including p$^+$ and n$^+$ contact regions), and achieves more than 50% improvement in comparison with the reference diodes thanks to the optical reflection from the bottom mirror. Only 2%–12% variations of responsivities are observed for the on-membrane diodes when the operation temperature increases up to 200°C, demonstrating the device capability for high-temperature photodetection.

Combined with the previously described thermal sensing capability of diodes under forward bias and using the microheater in the platform as a heat source, the suspended SOI PIN photodiode (when operated under reverse bias) can reliably achieve improved optical response up to a high temperature of about 200°C with *in situ* temperature sensing and control.

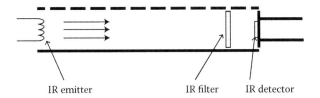

FIGURE 5.9
Schematic representation of a NDIR sensor system: a gas-permeable tube with an IR emitter and detector at each end and a band pass filter to restrict the radiation wavelength reaching the detector.

5.2.6 CO_2 Sensor

The most well-known method to measure carbon dioxide (CO_2) concentration is through the use of the non-dispersive-infra-red (NDIR) technique [27,28]. This method relies on the fact that CO_2 absorbs IR radiation near the 4.26 μm wavelength. Figure 5.9 shows the working principle of a NDIR sensor. A tube has a broadband IR emitter at one end and an IR detector at the other. A band pass filter at 4.26 μm only allows radiation near this wavelength to reach the detector. The tube has one or more holes to allow gas diffusion. As the CO_2 concentration in the tube changes, the amount of IR radiation absorbed at 4.26 μm changes, and so too does the signal at the detector, which is used to determine the CO_2 concentration.

Measuring CO_2 at high temperatures has a few challenges. All the devices and the tube should be able to withstand high temperatures. For example, the tube surface, which often has reflections, needs to not deteriorate over time. The IR emitter will also give a lower signal, since the temperature difference it generates is smaller. For instance, if at RT the IR emitter was pulsing from 25°C to 500°C, now it is pulsing from 225°C to 500°C, which results in a lower peak-to-peak voltage at the detector.

Fortunately, the CO_2 concentration in the exhaust of a domestic boiler is between 6% and 14% by volume, much higher than ambient air concentrations (~400 ppm), so the lower signal on the detector is not so problematic. Additionally, the high concentration requires a very small distance between the emitter and detector (typically 5 mm), so the tube reflectance is not critical, as the radiation path is almost direct between the emitter and the detector.

A large microhotplate device was used as an IR emitter. This device had a 1.8 mm × 1.8 mm heater (Figure 5.10), and at 450°C consumes 400 mW of power in direct current (DC) operation. It was packaged with a high-temperature die attach to withstand up to 225°C. A commercial IR detector (Heimann J21, with a 4.26 μm filter and half-pass bandwidth of 90 nm) was used. The detector and emitter were placed at opposite ends of an aluminium tube. A high-temperature printed circuit board (PCB) with a 4000×

FIGURE 5.10
Photo of an IR emitter chip packaged onto a TO39 package.

FIGURE 5.11
Photo of an IR detector chip on a high-temperature PCB.

amplifier circuit was connected with the detector, as shown in Figure 5.11. The amplifier circuit needs to be right next to the detector, as the signal detected is very small. The tube was placed in a small chamber to allow gas flow.

The setup was placed in an oven heated at 225°C. The IR emitter was pulsed at 1 Hz, and the signal at the detector was recorded. The peak-to-peak value of each pulse gives the IR radiation value. Different concentrations of CO_2 were passed through the system, and the peak-to-peak detector signal was recorded. The signal was filtered by averaging over 30 s and is shown in Figure 5.12, showing sufficient sensitivity and accuracy for boiler applications.

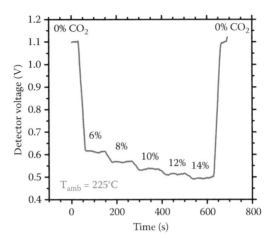

FIGURE 5.12
Response of the NDIR sensor to CO_2 at 225°C.

5.3 Packaging

There is pressure within a wide range of high-temperature applications to move the electronics closer to sensors. The closer proximity of the electronics allows improved efficiency through more rapid feedback and reduced weight of interconnecting pipework and cabling. In essence, in case of combustion control applications, the closer the electronics are to the point of combustion, the better the ability to control the efficiency of burn.

However, substrates, interconnections and assembly methods for commercial applications are not currently adequate for temperatures above around 125°C. Current harsh environment electronics must use technologies that date back to the 1970s; for example, high-temperature integrated circuits will typically use ceramic dual in-line-style packages. Further high-temperature circuit boards are often based on expensive ceramic processes such as low-temperature Co-fired ceramic (LTCC) or ceramic thick film. Both of these processes are subject to high tooling costs, and therefore limit take-up to high-volume or high-margin product.

In the early part of 2011, Microsemi did a top-level assessment of PCB assembly technologies intended for harsh environments. The output of this investigation identified that a new approach was required for substrates and assemblies. Under the SOI-HITS project, Microsemi began a study of lighter and cheaper materials suitable for use as alternatives to ceramic. However, developing the substrate is only one aspect of producing a 'system in package'; methods for soldering, die attach, component assembly and wiring were also evaluated.

Initial investigations involved the design of test patterns that represented a repeatable method for evaluating substrates and assemblies. But as confidence grew, more complex designs were developed, resulting in a full multichip module design.

Work on the low-cost organic packaging conducted by Microsemi has shown significant progress during the past 4 years. A complete assembly process that is compatible with PCB design and manufacturing was developed, and organic-based systems capable of reliable long-term operation at 225°C were demonstrated.

Accelerated ageing at high-temperature represents a significant problem in its own right because most testing then has to be done in near real time. Testing of the new technologies continues, but at the time of writing, Microsemi's high-temperature organic systems in package technology have been subjected to and passed the following reliability testing:

High-Temperature Organic Assembly Test Results	
Long-term exposure operation	10,000 h (>1 year) at 250°C
Thermal cycling of PCB (not assembly)	1,000 cycles –20°C to 250°C
Humidity exposure	1000 h, 85°C/85% humidity

Microsemi set out to develop a viable alternative to ceramic-based high-temperature assembly technology. This objective has been achieved, and now implementation of these technologies, including multilayer high-temperature PCBs and high-temperature lead-free assemblies, has started.

5.4 Sensors System

Integrating the harsh environment sensors with signal conditioning, drive and read-out circuitry on a single high-temperature PCB guarantees fast, low-noise and accurate measurements.

The drive and read-out circuitry is implemented as a high-temperature multisensor electrical interface using SOI technology to withstand high temperature and is used for high-precision signal conditioning and interfacing. This interface consists of two functional blocks: a read-out interface (ROI) chip providing ancillary circuits to the sensor chips (heater drivers, biasing and particular sensor interfaces) and a generic sensor interface (GSI) chip providing an analogue multiplexer, an analogue-to-digital converter with high-precision signal conditioning and a digital interface with a resolution of up to 10 bits, low noise and low offset. The outputs of the sensors are voltages and are read by the GSI chip, except for the capacitive humidity sensors whose output is converted to frequency by a ring oscillator within the ROI chip. The oxygen sensor operates typically at a constant temperature of

FIGURE 5.13
High-temperature PCB for harsh environment sensors and SOI circuitry.

500°C and is driven by the ROI chip using a pulse-width-modulated (PWM) technique. As the ambient temperature varies, the duty cycle and frequency of the PWM signal are varied to keep the operating temperature constant as the ambient temperature rises or falls.

The sensors and the associated circuitry are placed on a high-temperature PCB, shown in Figure 5.13. The PCB is made from a lower-cost organic substrate with gold-plated copper tracks providing working temperatures up to 280°C. It houses an oxygen and a temperature–humidity sensor, together with the ROI and GSI chips, and is interfaced using high-temperature paste and wires to a Freescale Freedom board containing a KL25Z 32-bit microcontroller.

Electrothermal characterisation of harsh environment sensors requires an automated test system that is capable of monitoring the performance of multiple sensors at different temperatures and RH (typically 20°C–400°C and 0.1%–100%, respectively), creating and delivering a precise mixture of target gases, and performing thermal cycling for long-term reliability testing.

The general configuration of such a smart gas testing system is show in Figure 5.14.

The analogue front end comprises the gas delivery subsystem using high-precision digital mass flow controllers and a gas test chamber containing the high-temperature PCB housing the gas sensors and their associated circuitry. For harsh environment characterisation, the gas test chamber is placed in a computer-controlled commercial electric oven (Memmert UNP-200).

In order to control the measurement setup in a synchronised and consistent way, all of the subsystems – sensor setup, gas delivery and high-temperature oven – are controlled by an integrated control and data acquisition user interface implemented in Labview software (National Instruments).

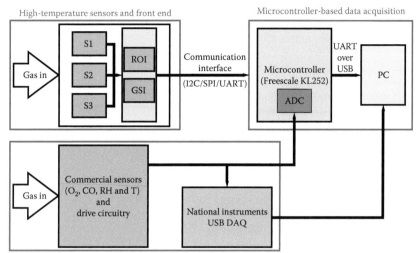

FIGURE 5.14
Schematic block diagram of the smart gas testing system.

A parallel system containing commercial sensors is added to confirm the functional equivalence of the high-temperature sensors. This benchmark system is driven by a National Instruments USB data acquisition module controlled by a laptop running Labview.

References

1. Schultz, M. 2011. Gas appliance modulating controls technology. Presented at American Society of Gas Engineers Conference, Las Vegas, NV, June 7.
2. Udrea, F., et al. 2008. Ultra-high temperature (>>300 °C) suspended thermo-diode in SOI CMOS technology. In 14th International Workshop on Thermal Investigation of ICs and Systems (THERMINIC 2008), Rome, September 24–26, pp. 195–199.
3. Santra, S., et al. 2010. Silicon on insulator diode temperature sensor – A detailed analysis for ultra-high-temperature operation. *Sensors Journal IEEE* 10 (5): 997–1003.
4. De Luca, A., et al. 2015. Experimental, analytical and numerical investigation of non-linearity of SOI diode temperature sensors at extreme temperatures. *Sensors and Actuators A* 222: 31–38.
5. Kuo, T. J. W., et al. 2012. Micromachined thermal flow sensors – A review. *Micromachines* 3 (3): 550–573.
6. Chen, B., et al. 2014. Effects of ambient humidity on a micromachined silicon thermal wind sensor. *Journal of Microelectromechanical Systems* 23 (2): 253–255.

7. De Luca, A., et al. 2015. SOI multidirectional thermoelectric flow sensor for harsh environment applications. In *2015 International Semiconductor Conference (CAS)*, Sinaia, Romania, October 14–16, 2013, pp. 95–98.

8. Kovac, M. G., et al. 1978. A new moisture sensor for in situ monitoring of sealed packages. *Solid State Technology* 21: 35–39.

9. Westcott, L., and Rogers, G. 1985. Humidity sensitive MIS structure. *Journal of Physics E: Scientific Instruments* 18: 577–586.

10. Shi, X., et al. 2008. Al2O3-coated microcantilevers for detection of moisture at ppm level. *Sensors and Actuators B* 129: 241–245.

11. Juhasz, L., et al. S. 2008. Porous alumina based capacitive MEMS RH sensor. In *Symposium on Design, Test, Integration and Packaging of MEMS/MOEMS*, Nice, France, April 9–11, pp. 381–385.

12. André, N., et al. 2015. Silicon-on-insulator micro-hotplates platforms for humidity sensing: Follow-up of the air quality intercomparison exercise. Presented at 3rd International Workshop of EuNetAir, Riga, Latvia, March 26–27.

13. Moos, R., et al. 2011. Resistive oxygen gas sensors for harsh environment. *Sensors* 11: 3439–3465.

14. Williams, D. E., et al. 1982. Oxygen sensors. European Patent 0,062,994.

15. Neri, G., et al. 2008. $FeSrTiO_3$-based resistive oxygen sensors for application in diesel engines. *Sensors and Actuators B* 134: 647–653.

16. Choi, S.-H., et al. 2013. Facile synthesis of p-type perovskite $SrTi_{0.65}Fe_{0.35}O_{3-\delta}$ nanofibers prepared by electrospinning and their oxygen-sensing properties. *Macromolecular Materials and Engineering* 298 (5): 521–527.

17. Cobianu, C., et al. 2013. Iron doped strontium titanate powder. European Patent Application EP 13184471.4. Filed September 14.

18. Suslick, K. S. 1997. Sonocatalysis. In *Handbook of Heterogeneous Catalysis*, ed. G. Ertl, H. Knozinger, and J. Weitkamp. 1st ed., Vol. 3. Weinheim: Germany, pp. 1350–1357.

19. Cobianu, C., et al. 2013. Iron doped strontium titanate powder. European Patent Application EP 13196767.1. Filed December 11.

20. Stratulat, A., et al. 2015. Low power resistive oxygen sensor based on sonochemical $SrTi_{0.6}Fe_{0.4}O_{2.8}$ (STFO40). *Sensors* 15: 17495–17506.

21. Neri, G., et al. 2007. Resistive λ-sensors based on ball milled Fe-doped $SrTiO_3$ nanopowders obtained by self-propagating high temperature synthesis (SHS). *Sensors and Actuators B* 126: 258–265.

22. Ginger, D. S., et al. 2004. The evolution of dip-pen nanolithography. *Angewandte Chemie International Edition* 43: 30–45.

23. Stratulat, A., et al. 2015. Novel sonochemical route for manufacturing O_2 sensitive STFO. In *NATO Advanced Research Workshop: Functional Nanomaterials and Devices for Electronics, Sensors Energy Harvesting*, Lviv, Ukraine, April 13–16, 2015, pp. 78–79.

24. Afzalian, A., and Flandre, D. 2005. Physical modeling and design of thin-film SOI lateral PIN photodiodes. *IEEE Transactions on Electron Devices* 52 (6): 1116–1122.

25. Andre, N., and Li., G. L. 2015. Wide band study of silicon-on-insulator photodiodes on suspended micro-hotplates platforms. In *17th International Conference on Integrated Circuit Design and Technology*, Leuven, Belgium, June 1–3, pp. 1–4.

26. Bulteel, O. 2011. Silicon-on-insulator optoelectronic components for micro-power solar energy harvesting and bio-environmental instrumentation. PhD dissertation, Université Catholique de Louvain, Louvain-la-Neuve, Belgium, pp. 16–18.
27. Hodgkinson, J., and Tatam, R. P. 2013. Optical gas sensing: A review. *Measurement Science & Technology* 24: 012004.
28. Frodl, R., and Tille, T. 2006. A high-precision NDIR CO_2 gas sensor for automotive applications. *IEEE Sensors Journal* 6: 1697–1705.
29. Kroetz, G. H., et al. 1999. Silicon compatible materials for harsh environment sensors. *Sensors and Actuators A* 74: 182–189.
30. Cimalla, V., et al. 2007. Group III nitride and SiC based MEMS and NEMS: Materials properties, technology and applications. *Journal of Physics D* 40: 6386–6434.
31. COTS2.21. Commercial humidity sensor HCH 1000. Available at http://datasheet.octopart.com/HCH-1000-001-honeywell-datasheet-9695259.pdf.
32. COTS2.22. Commercial humidity sensor SHT21. Available at www.sensirion.com.
33. COTS2.23. Commercial humidity sensor MK33. Available at www.ist-ag.com.
34. Kim, J. S., et al. 2009. Fabrication of high-speed polyimide-based humidity sensor using anisotropic and isotropic etching with ICP. *Thin Film Solid*, 517: 3879–3882.
35. Dai, C. L. 2007. A capacitive humidity sensor integrated with micro heater and ring oscillator circuit fabricated by CMOS–MEMS technique. *Sensors and Actuators B* 122: 375–380.

6

III-Nitride Electronic Devices for Harsh Environments

Shyh-Chiang Shen

CONTENTS

ABSTRACT III-nitride (III-N) devices are feasible for next-generation harsh environment applications owing to their unique material properties. With low resistive loss and suitability for high-voltage operation, III-N devices are actively being developed for high-power microwave amplifications and high-voltage power switching with operating temperatures beyond 200°C. These harsh environment applications include downhole drilling, electric vehicles, on-board engine control systems and the outdoor power grid. This chapter provides an introductory review of the development of III-N electronic devices. A brief overview of the material fundamentals of III-N semiconductors and design considerations for polar semiconductor devices is presented. State-of-the-art III-N high-electron-mobility transistors (HEMTs) for high-power microwave/millimetre-wave and high-voltage switching, as well as novel III-N bipolar switches, including heterojunction bipolar transistors (HBTs) and rectifiers, are discussed to exploit the potential of III-N devices in ultra-high-power and high-temperature applications. Finally, a summary of the radiation effect study on III-N materials and devices is presented.

6.1 III-Nitride Semiconductor Properties

III-nitride (III-N) materials offer the flexibility of using various compositions of Group III elements (e.g. In, Al and Ga) and the nitrogen in a semiconductor system, which has great potential for next-generation harsh environment applications. Commonly used III-N semiconductors include gallium nitride (GaN), aluminium nitride (AlN), indium nitride (InN) and their ternary and quaternary alloys, such as InGaN, InAlN, AlGaN and AlInGaN. They are known to have direct-energy bandgap, comparable thermal conductivity, chemical inertness and high electron mobility.

Table 6.1 lists major physical properties of common semiconductors. Unlike elemental semiconductors, many compound semiconductors possess the direct-energy bandgap property and have been favourably used for efficient light emitters, such as light-emitting diodes (LEDs) and laser diodes (LDs). Complementing the extensive use of 'low-bandgap semiconductors' (InP and GaAs-based compound semiconductors) in the infrared (IR) and the green to red wavelengths, the III-N material development, specifically in the realisation of high-quality-doped III-N materials and the related (Al)(In) GaN/(In)GaN quantum-mechanical heterostructures, has served as a technological basis for today's energy-efficient solid-state lighting business. The bandgap engineering in the III-N materials system also facilitates the realisation of compact ultraviolet (UV) light sources in the near UV to the deep-UV wavelengths, which could be an important technology for water sanitisation systems and next-generation photolithography tools.

TABLE 6.1

Material Properties of Selected Semiconductors at 300 K

	Binary III-N			Si	Ge	GaAs	InP	4H-SiC
	GaN	AlN	InN					
Crystalline	W	W	W	Di	Di	Z	Z	W
Band structure	D	D	D	I	I	D	D	I
a (Å)	3.189	3.11	3.544	5.431	5.646	5.653	5.869	3.073
E_g (eV)	3.44	6.2	0.7	1.12	0.66	1.42	1.35	3.26
E_{cr} (MV/cm)	~5	~1.8	—	0.8	0.1	0.3–0.9	0.5	3–5
$v_{e,sat}$ (10^7 cm/s)	2.5	1.9	3.4	1.0	3.1	0.7	3.9	1.9
μ_n (cm^2/V·s)	1000	300	3200	1400	3900	8000	5400	900
μ_p (cm^2/V·s)	400	14	—	500	1900	400	200	120
κ (W·cm^{-1}K^{-1})	2.0–2.4	3.0–3.3	0.6–1.0	1.56	0.58	0.46	0.68	3.7

Source: After Zhang, Y., PhD dissertation, Development of III-Nitride Bipolar Devices: Avalanche Photodiodes, Laser Diodes, and Double-Heterojunction Bipolar Transistors, Georgia Institute of Technology, 2011.

Note: W = würtzite, Di = diamond, Z = zinc blende; D = direct-bandgap semiconductor, I = indirect-bandgap semiconductor, a = lattice constant, E_g = bandgap energy, E_{cr} = critical electric field, $v_{e,sat}$ = electron saturation velocity, μ_n = electron mobility (for intrinsic or unintentionally doped material), μ_p = hole mobility (for intrinsic or unintentionally doped material), κ = thermal conductivity.

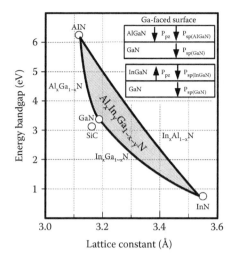

FIGURE 6.1
Bandgap energy vs. lattice constant of the III-N materials system. The inset indicates the polarisation alignments for AlGaN/GaN heterojunctions grown on a Ga-faced substrate.

The relationship of the energy bandgap (E_g) and the lattice constant (a_o) of a III-N semiconductor in the form of the würtzite crystalline is shown in Figure 6.1. The terminal binary compounds of interests are AlN, GaN and InN. The curves connecting two terminal binary compounds represent the E_g versus a_o (E_g-a_o) relationship of a ternary material. For example, the curve connecting AlN and GaN is the E_g-a_o relationship of an Al$_x$Ga$_{1-x}$N compound, where the lattice constant can be determined by Vegard's law and the energy bandgap can be modelled using a composition-dependent polynomial. The shaded area represents the E_g-a_o relationship of a quaternary alloy, Al$_x$In$_y$Ga$_{1-x-y}$N ($0 < x,y < 1$). In comparison to elemental semiconductors (e.g. Si or Ge), III-N semiconductors could cover a wide range of bandgap energy from 0.7 eV (for InN) to 6.2 eV (for AlN). Typically, GaN, InAlN and AlGaN alloys are suitable choices for electronic applications that can operate under high-temperature, high-power and radioactive environments.

The bandgap engineering in III-N materials facilitates the design of advanced electronic devices. Devices such as heterojunction bipolar transistors (HBTs) and high-electron-mobility transistors utilise the advantages of the device concepts from conventional III–V semiconductor fields and leverage their unique wide-bandgap (WBG) properties for harsh environment applications. It is noted that another WBG material, silicon carbide (SiC), which can be found in today's high-temperature and high-power switching circuitry, does not have the bandgap engineering versatility. In addition to a high critical electric field, III-N materials have higher electron velocity by a factor of at least two when compared with silicon, and are favourable for high-speed and high-power circuits. The direct-bandgap III-N semiconductors have a short carrier recombination lifetime in the range of a few nanoseconds, which may lead to fast switching transient in energy-efficient power electronic systems. III-N materials also have good thermal conductivity for high-power operations.

Advanced III-N materials can be epitaxially grown in commercially available reactors. The commonly used epitaxy methods are molecular beam epitaxy (MBE), metal-organic chemical vapour deposition (MOCVD), and hydride vapour phase epitaxy (HVPE). By sequencing proper sets of chemical precursors in a reactor, III-N heterostructures are formed by the monolithic growth of semiconductor layers of designated alloy compositions with atomic-level accuracy. III-N heterostructures usually come at the cost of internal mechanical stress build-up due to the lattice mismatch between epitaxial layers. The impacts of the lattice mismatch are at least threefold. First, the strain in a heterostructure limits the maximal thickness of each layer before it is structurally relaxed, known as the critical thickness. A relaxed epitaxial layer tends to create crystalline dislocations and cracks in these semiconductor structures. Second, a strained semiconductor layer alters the deformation potential and changes the electron and hole transport properties accordingly. Third, the strain will significantly change the polarisation state in the semiconductors. Compared with cubic semiconductors, III-N materials typically form in the würtzite crystalline structure (see Table 6.1). It consists of two interpenetrating hexagonally closely packed basal planes, one with Group III elements only and the other with Group V elements only. Each of the Group III elements is tetrahedrally coordinated with the nearest four Group V elements. With a lack of inversion symmetry, III-N materials in the würtzite form have a large polarisation charge and piezoelectric coefficient along the c-axis of the crystalline. The polarisation-related parameters for AlN, GaN and InN are summarised in Table 6.2.

The lattice mismatch in a heterostructure creates the lattice deformation that results in the piezoelectric effect in III-N materials. Take an epitaxial layer grown on gallium-faced GaN for an example. If a strained AlGaN layer is grown on top of a GaN layer, it will experience the tensile strain. The elastically strained AlGaN/GaN material will experience a piezoelectric field along the c-axis (P_{pz}) induced through the mechanoelectric transduction process.

TABLE 6.2

List of the Lattice Constant, Spontaneous Polarisation Charge, Piezoelectric Coefficient and Stiffness Constant for AlN, GaN and InN

	GaN	AlN	InN	Reference
a_0 (Å)	3.189	3.112	3.54	73
P_{sp} (C/m²)	−0.034	−0.090	−0.042	74
e_{31} (C/m²)	−0.49	−0.6	−0.57	74
e_{33} (C/m²)	0.73	1.46	0.97	74
C_{13} (GPa)	103	108	92	75
C_{33} (GPa)	405	373	224	75

This P_{pz} has a polarisation field lined up with the spontaneous polarisation field (P_{sp}), leading to an enhanced polarisation effect, as shown in the top inset of Figure 6.1. On the other hand, if a compressively strained III-N material (e.g. InGaN) is grown on a GaN layer, P_{pz} has a field that is lined up in the opposite direction to P_{sp} and the net effect is a reduced overall polarisation. If a semiconductor structure is grown on a nitrogen face of GaN, the polarisation field line-up will be reversed accordingly; if a semiconductor structure is grown on nonpolar (*m*-) or semipolar planes, the semiconductor layers along this growth direction will be less susceptible or free from the polarisation field effect.

The presence of the polarisation charges and the associated polarisation field leads to unique design strategies for III-N-based devices. For example, a two-dimensional electron gas (2DEG) channel may be induced solely by polarisation charges in an undoped heterointerface, leading to reduced impurity scattering and high sheet charge density in III-N heterojunction field-effect transistors (HFETs). On the other hand, the built-in polarisation field may give rise to significant band profile tilting that affects the electron and hole transport and skews the wavefunction distribution. Due to the lack of low-cost native substrates, III-N devices are commonly grown on foreign substrates such as sapphire, silicon carbide and silicon. Engineered buffer layers are usually inserted in between the substrate and the active device layers for strain management purposes. The interplay between strain management, polarisation field engineering, substrate orientation and impurity doping schemes would need careful scrutiny for successful III-N device implementation.

Although current silicon-based electronics, including SiGe electronics, cover a wide range of temperature operation from the cryogenic state up to ~200°C, circuits built on low-bandgap semiconductors become impractical for higher-temperature applications due to a significantly increased concentration of thermally generated electron–hole pairs. With more than two decades of active development, III-N devices have evolved from academic research interests to commercial products. Since III-N materials have low intrinsic carrier concentration, III-N devices have become a suitable choice for high-temperature operation. III-N devices may operate at high junction temperature with deferred thermal-induced device degradation. They can also operate at high ambient temperature where silicon or GaAs electronics cease to perform properly.

6.2 III-N FETs

6.2.1 III-N FET Design and Fabrication

The conceptualisation of III-N HFETs was derived from the conduction modulation of the 2DEG in GaAs- and InP-based HEMTs. When compared with conventional III–V HEMTs, however, today's III-N HEMTs have a relatively simple epitaxial layer design. It was shown that the polarisation-induced electric field alone can lead to a significant amount of free carrier accumulation at the AlGaN/GaN interface even without the inclusion of a modulation-doped layer [1]. With proper selections of crystalline orientation, thicknesses and compositions of III-N materials, either the 2DEG or the two-dimensional hole gas (2DHG) can be induced at the heterointerface of intrinsic semiconductor layers. This unique feature is neither observed nor possible for conventional HEMTs in cubic semiconductors. As a result, an effective III-N HEMT structure may be composed of a thin GaN layer with a thickness of a few microns and an AlGaN layer with a thickness of a few tens of nanometres to form an AlGaN/GaN HEMT. Most III-N HFETs to date are grown on a Ga-polar surface to leverage the polarisation-induced 2DEG at the AlGaN/GaN heterointerface. Other heterojunction designs include AlGaN/AlN/GaN HFETs and InAlN/GaN HFETs [2–4]. These devices can be implemented on a wide variety of substrate platforms, such as sapphire, silicon carbide, free-standing (FS) GaN or (111) Si. Impressive high-power microwave performance has been reported in each substrate platform by either research laboratories or commercial companies. III-N HEMTs can also be grown on the nitrogen-polar (N-polar) surface to take advantage of a natural back-barrier to form a double heterojunction for effective 2DEG confinement. However, high-quality epitaxial growth of N-polar III-N materials has been challenging. It was not until recent years that one could see a demonstration using N-polar GaN/AlGaN HFETs [5].

The fabrication processing of III-N HEMTs or HFETs is comparable to conventional compound semiconductors except for slight modifications in the ohmic contact formation, semiconductor etching chemistry and device passivation. The threshold voltage control schemes for III-N FET devices were actively explored. For example, a recessed-gate structure can be implemented by either a wet-etching or dry-etching process [6]; a high-k dielectric layer was deposited in the gate region to form metal-insulator-semiconductor (MIS) HEMTs [7,8]; a p-n junction was embedded in the gate region to emulate a junction-gate field-effect transistor (JFET)–like 'gate-injected transistor' for normally-off operation [9]; and plasma surface treatments such as oxidation [10] or fluorine incorporation [11] are also investigated. It has been reported that a GaN-based HEMT could operate at temperatures as high as 1000°C [12]. However, the high-temperature characteristics vary from one particular device design to another, in terms of the gate structures (Schottky, MIS, p-n junction, etc.), passivation methods (benzocyclobutene, spin-on glass, silicon nitride, etc.) and ohmic contact schemes. The performance of each type of device under high-temperature environment operation would need further study on a case-by-case basis.

6.2.2 High-Power Microwave and Millimetre-Wave Performance

AlGaN/GaN HEMTs have demonstrated impressive radio frequency (RF) power amplification characteristics, and GaN HEMT-based power amplifier modules are commercially available [13–20]. For example, AlGaN/GaN HFETs that were grown on a SiC substrate can achieve a maximal drain current density ($I_{D,max}$) of 1.2 A/mm, a unit-gain cut-off frequency (f_T) of 70 GHz and a unilateral power gain frequency (f_{max}) of 300 GHz [21]; AlGaN/GaN HFETs grown on a silicon substrate also demonstrated good performance and achieved an $I_{D,max}$ of 800 mA/mm, f_T of 100 GHz and f_{max} of 206 GHz [22]. Through aggressive device scaling schemes and regrown drain and source, state-of-the-art III-N HEMTs have achieved impressive f_T and f_{max} values of 310 and 364 GHz, respectively [23]. For RF power amplification, III-N HEMTs showed a high-output power density of greater than 10 W/mm at the Ka-band [24] and demonstrated W-band power amplification with an RF density of greater than 2 W/mm at 80 GHz [25]. For L-band applications, III-N HEMTs also showed an impressive output power of 500 W [26]. The high-power RF amplification performance was attributed to the extended breakdown field and high saturation electron velocity characteristics in III-N materials.

6.2.3 High-Voltage Switching Performance of III-N FETs

The potential of AlGaN/GaN HFETs for high-voltage electronics has also received much attention in recent years. It was reported that AlGaN/GaN HEMTs can operate at a blocking voltage greater than 1 kV with the on-state resistance surpassing the 6H-SiC limit [27–32]. Various III-N FET designs, including trenched gate [33], slant field plate [34], source field plate [35] and reduced surface field (RESURF) metal oxide semiconductor field-effect transistors (MOSFETs) [36], were reported. In vertical devices, III-N transistors using vertical electron conduction [37], vertical insulated gates [38] or U-shaped trenched gates [39] were also studied. AlGaN/GaN HFETs with a blocking voltage of 8.3 kV and an on-state resistance of 168 mΩ·cm² were successfully demonstrated [35]. These results indicate that III-N transistors could be a suitable device technology for power electronic components with a blocking voltage rating of 600 V and beyond.

As an example, a high-voltage normally-off AlGaN/AlN/GaN HFET technology, which used a recessed-gate structure, was developed [40]. The device structure was grown on a high-resistivity (111) silicon substrate, and a schematic cross-sectional view of the device is shown in Figure 6.2. The fabrication process starts with device mesa isolation in an inductive coupling plasma (ICP) etching tool, followed by a KOH-based recessed-gate etching step. The AlN layer serves as a polarisation charge enhancement layer, as well as an etch-stop layer for the recessed-gate process [41]. Ti/Al-based

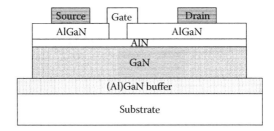

FIGURE 6.2
Schematic cross-sectional view of a recessed-gate AlGaN/AlN/GaN HFET. The substrate of choices can be (111) silicon, silicon carbide, sapphire or FS-GaN.

ohmic metal contact was formed in the drain and the source region, and Ni/Au was used for the Schottky gate metal. The HFETs were passivated by benzocyclobutene. A 1 μm thick Ti/Au interconnect metal layer was deposited after via holes were opened on the passivated devices. For a comparison, devices with as-grown and gate-recessed structures, respectively, were fabricated and evaluated on the same wafer. The measured I_D-V_{GS} transfer curve of a 0.3 mm wide HFET is shown in Figure 6.3a at $V_{DS} = 10$ V. The device has a gate-to-drain distance (L_{GD}) of 13 μm and a gate length (L_G) of 3 μm. The threshold voltage (V_{th}), determined at $I_{DS} = 1$ mA/mm, is –6 V for the as-grown HFET. V_{th} is shifted to 0.06 V for the recessed-gate HFET. The on–off ratio is approximately identical (2×10^6) on either the as-grown or the gate-recessed devices. The maximum transconductance ($g_{m,max}$) is increased from 86 mS/mm for the as-grown HFET to 116 mS/mm for the gate-recessed HFET due to the reduced barrier thickness in the gate region. The off-state drain leakage current remains <200 nA/mm for devices with and without the recessed-gate etching. The results confirm that good Schottky gate properties can be achieved on recessed-gate HFETs without inducing additional etching damage through the recessed etching process.

In Figure 6.3b, the I_D-V_{DS} family curves of the as-grown HFET show $I_{D,max}$ values of greater than 510 mA/mm at $V_{GS} = 1$ V. On the other hand, the recessed-gate device shows lower $I_{D,max}$ (420 mA/mm) at $V_{GS} = 4$ V. The specific on-resistance ($R_{DS(on)}$) is 4.7 mΩ-cm² for the as-grown HFET (normally-on device), and that for the gate-recessed HFET (normally-off device) is 6.6 mΩ-cm². Higher on-state resistance and lower maximum drain current in the normally-off III-N HFETs are attributed to increased channel resistance in the recessed-gate region as a trade-off design between V_{th} and $R_{DS(on)}$ in these devices. In terms of high-voltage operation, the normally-off HFET with $L_{GD} = 13$ μm showed a breakdown voltage of greater than 1.2 kV at $V_{GS} = -1$ V, as shown in Figure 6.3c. The corresponding lateral breakdown field is 0.92 MV/cm.

The research and development of III-N HFETs for high-voltage switching encompasses all fronts of activity, from substrate preparation, heterogeneous epitaxial material growth and device processing to power electronic circuit

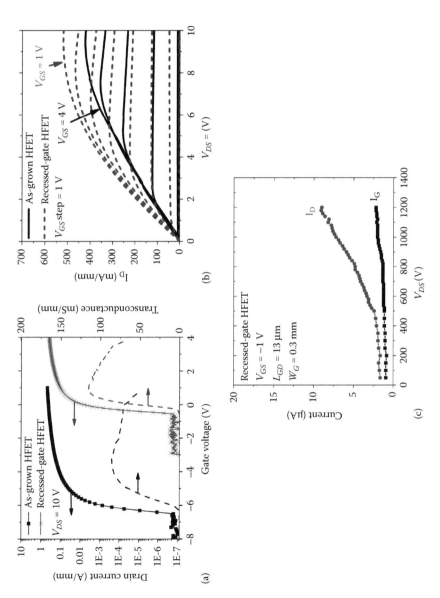

FIGURE 6.3

(a) Measured I_D-V_{GS} transfer curves. (b) I_D-V_{DS} family curves of a 0.3 mm wide AlGaN/GaN HFET with and without recessed-gate etching. (c) Drain and gate leakage current of a recessed-gate AlGaN/GaN HFET with $W_G = 0.3$ mm and $L_{GD} = 13$ μm at $V_{GS} = -1$ V.

demonstration and system integration. Although several potentially show-stopping device performance issues, such as dynamic on-state resistance, threshold voltage stability and surface-state-related hysteresis effect, have not been fully resolved, several companies have begun offering engineering sampling and commercial products. As of 2015, commercial GaN-based HFETs are available in the low-voltage end of direct current (DC)–DC converter circuits with a power device rating below 200 V. For example, a normally-off GaN FET could be used for a DC-to-DC downconversion architecture from 48 V to the point of load in server racks [42]. When compared with the circuits made with state-of-the-art Si power MOSFETs, the use of III-N HFETs in these converters improves the hard switching figure of merit by a factor of at least five for devices with a blocking voltage below 200 V. The power GaN FET circuits demonstrated higher conversion efficiency by a few per cent at high-current states. With improved power conversion efficiency using GaN FETs, a much lower operation temperature at the circuit board level can be achieved, when compared with Si-based counterparts. Sampling of III-N HFETs for 600 V and 1.2 kV applications was also available as of 2015.

6.3 III-N Bipolar Transistors

6.3.1 Development of High-Current-Gain, High-Power-Density III-N HBTs

III-N HBTs have long been anticipated to follow the paths of GaAs- and InP-based HBTs and offer a new class of transistor technology for next-generation microelectronic systems. HBTs, in general, have superior wafer-level uniformity in the bandgap-dictated turn-on voltage and high transconductance for efficient RF power amplification. The normally-off characteristics are highly desirable for single-power supply circuits. Due to the nature of vertical carrier transport, HBTs are less susceptible to the surface-state-induced leakage and can achieve higher operating power density and low on-state resistance. When the device is moderately modified, it may include a surface conduction path, such as a gate-controlled electrode, to implement insulated-gate bipolar transistors (IGBTs) in III-N materials for future energy-efficient switches and high-power electronics.

To date, the III-N HBT technology is not as well developed as the III-N HEMT technology. Issues stemming from material growth techniques and device fabrication processing were less explored. III-N HBTs in the *npn* configuration are a particular challenge: the free-hole concentration of the Mg-doped *p*-type III-N layers is limited due to fundamental issues related to the deep acceptor state, the lattice mismatch between epitaxial layers and the use of foreign substrates prone to threading dislocations and stacking faults, and the dry-etching-induced nitrogen-rich surface leads to high *p*-type base contact resistance, to name a few of the primary issues.

The 'first-generation' III-N HBTs that can achieve high-current and high-gain characteristics were implemented in *npn* GaN/InGaN heterojunctions because a reasonably high free-hole concentration and slow surface recombination velocity can be achieved in a *p*-type InGaN layer. In early *npn* GaN/InGaN HBT development, not only was the InGaN base layer epitaxially grown to reduce the *p*-type base sheet resistance, but also an extrinsic *p*-InGaN base regrowth process was employed to reduce the base resistance of an *npn* GaN/InGaN HBT. Through multiple growth steps, Makimoto et al. were able to demonstrate *npn* GaN/InGaN HBTs with a current gain of greater than 2000 and a DC power handling capability of 270 kW/cm² [43,44]. The *pnp* InGaN HBTs were also demonstrated, with a good current drive capability [45]. It was found that GaN/In$_x$Ga$_{1-x}$N ($x = 0.09$) HBTs grown on SiC substrates were fully strained, while those grown on sapphire substrates were relaxed [46]. To successfully demonstrate III-N HBTs, Chung et al. and Dupuis et al. at the Georgia Institute of Technology carefully engineered the MOCVD growth and achieved high-quality HBT epitaxy using graded double heterojunctions to get around the growth defect formation in III-N HBTs [47,48].

To reduce the fabrication complexity, the Georgia Tech team also developed a single-pass growth (direct-growth) *npn* GaN/InGaN HBT processing technology that led to significant reduction in the surface leakage [49] and successful demonstrations of GaN/In$_x$Ga$_{1-x}$N ($x = 0.03$) HBTs on a sapphire substrate with good current gain ($\beta > 100$) and a high-breakdown voltage of greater than >100 V [50]. With an effort to downscale the InGaN HBTs, the team also demonstrated the first microwave amplification characteristics with $f_T > 5$ GHz in III-N HBTs [51]. To date, the microwave performance of III-N HBTs has been improved to $f_T = 8$ GHz [52].

Since HBTs handle much higher current density than HFETs, the thermal conductivity of the substrate is also a concern. For GaN/InGaN HBTs grown on sapphire substrates, a significant self-heating effect arising from the poor thermal conductivity of sapphire has limited the highest achievable collector current density (J_C) to approximately 20 kA/cm² [53]. Low-defect pseudomorphically grown HBTs on native substrates would be necessary to further improve the device performance. GaN/InGaN HBTs built on FS-GaN substrates have shown a record J_C of greater than 120 kA/cm², an ultra-high-power handling capability of >3 MW/cm² and a peak breakdown field of >2.5 MV/cm [54].

6.3.2 High-Temperature Performance of III-N HBTs

Bipolar transistors usually experience a drastically reduced current gain and significant leakage current due to enhanced trap-assisted generation processes at an elevated temperature. High-temperature operation of III-N HBTs was studied and demonstrated. For example, a *pnp* AlGaN/GaN HBT was fabricated using a superlattice emitter design and was evaluated up to 590°C [55]. The peak common-emitter current gain (β_{max}) was 38, 26, and 3 at 50°C, 300°C and 590°C, respectively, on a device with an emitter size of 30×50 µm². The open-base common-emitter leakage current increases only by a factor of nine from 50°C to 590°C at a collector-emitter voltage (V_{CE}) of 40 V. The high-temperature characteristics of GaN/InGaN HBTs grown on sapphire were also reported up to 300°C [56]. The results showed that β_{max} of an *npn* GaN/InGaN HBT with an emitter area of 50×50 µm² changes from 20 at 25°C to 10 at 300°C. Due to the enhanced thermal activation of the *p*-type dopant in the base, the offset voltage was improved from >5 V at room temperature to ~2 V at 300°C. Further improvement of GaN/InGaN HBTs was achieved by growing such a heterostructure on low-defect-density substrates such as FS-GaN, and the high-temperature operation of *npn* GaN/InGaN HBTs was also studied [54]. Shown in Figure 6.4a is the Gummel plot of a fabricated GaN/InGaN DHBT grown on a FS-GaN substrate with an emitter size of 40×40 µm². With a drastic reduction of the dislocation defect density in FS-GaN and better lattice matching in epitaxial layers, the device shows an impressive β_{max} of 95 at 25°C and 35 at 250°C, respectively. It is worth noting that although the base current is increased at elevated temperatures, appreciable J_C values of greater than 2 kA/cm² can be achieved at elevated temperatures up to 250°C. The reduced β_{max} is attributed to the increased recombination rate and reduced emitter injection efficiency as the temperature increases. The reduced base resistance at elevated temperature was confirmed from the sheet resistance and C-V measurements. A higher free-hole concentration was measured at elevated temperature, as a result of a higher thermal activation rate in the Mg-doped InGaN base layer. The common-emitter family curves of the HBT are also shown in Figure 6.4b for temperatures at 25°C, 150°C and 250°C, respectively. The offset voltage is reduced from 0.8 V at 25°C to 0.3 V at 250°C, and the knee voltage from 5.2 to 2.75 V at $I_B = 500$ µA as the temperature increases from 25°C to 250°C. The open-base emitter breakdown characteristics showed that the open-base common-emitter breakdown voltage (BV_{CEO}) increases from 90 to 157 V as the temperature increases from 25°C to 250°C. The positive temperature coefficient for BV_{CEO} indicates that the impact ionisation process is the major breakdown mechanism for GaN/InGaN DHBTs.

The demonstrated GaN/InGaN HBTs at room temperature and high-temperature regions showcase the state-of-the-art high-power and elevated-temperature performance of III-N HBTs. They provide a strong support for the feasibility of achieving the high-quality junction property with well-controlled mesa-type bipolar transistor fabrication processing for the realisation of high-performance III-N bipolar transistors. The avalanche capability

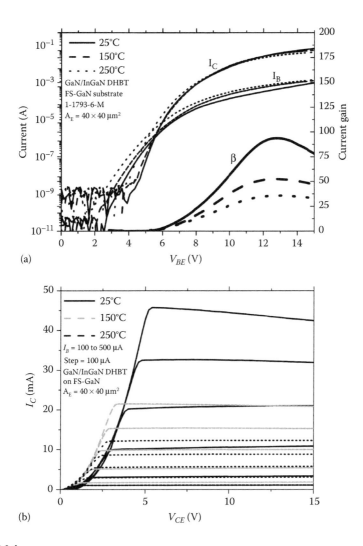

FIGURE 6.4

(a) Gummel plot. (b) Common-emitter family curves of a GaN/InGaN HBT grown on a FS-GaN substrate at 25°C, 150°C, and 250°C, respectively.

of III-N HBTs is a distinct and favourable feature when compared with III-N HFETs, as the catastrophic breakdown prevails in today's III-N HFETs, limiting further applications of III-N HFETs in power electronic systems. Furthermore, HBTs conduct the power flow in a vertical fashion, which is also a desired device form when it comes to high-voltage applications. For lower-voltage systems, the near-breakdown avalanche capability in III-N HBTs is also desired and could provide further improvement in energy efficiency in high-temperature electronics in the future.

6.4 III-N Rectifiers

III-N power rectifiers studied to date include *p-i-n* (PIN) diodes and Schottky barrier diodes (SBDs). Taking advantages of the 2DEG formation in the III-N heterojunctions, lateral Schottky diodes can also be implemented by appropriately modifying the device layout of the gate–source diodes in III-N HFETs. These rectifiers have attracted much research interest in recent years because of their low-conduction-loss, high-voltage and high-temperature operations for voltage regulators and DC-DC converters. When a nonconducting substrate is used, the anode and cathode of III-N rectifiers must be fabricated on the top side of the wafer, leading to additional on-state resistance and premature edge breakdown [57,58]. Recently, the availability of sufficiently large FS-GaN conducting substrates (>2 in. diameter) with a threading dislocation density of less than 10^6 cm^{-2} has helped the realisation of high-quality III-N epitaxy and high-voltage III-N rectifiers in a vertical current conduction form.

Vertical GaN SBDs have been reported with a turn-on voltage of less than 1 V with decent breakdown voltage [59,60]. GaN PIN rectifiers were also investigated for ultra-high-voltage switches because they possess several advantages over SBDs, such as ultralow leakage current and avalanche capabilities. Vertical homojunction GaN PIN rectifiers have shown impressive blocking voltages of greater than 3 kV and achieved >3 GW/cm^2 of switching figures of merit when they were built on FS-GaN substrates [61–63]. The challenges of the high-voltage PIN rectifiers, similar to III-N HBT development, are mostly related to the material growth. To maximise the breakdown voltage for a given thickness of the drift layer, one would need to minimise unintentional background doping during the material growth, in addition to proper field termination schemes such as field plate or tapered sidewall formation.

Laterally conducting SBDs that utilised the AlGaN/GaN 2DEG channel were reported to be operable for temperatures up to 175°C [64]. Similar devices grown on sapphire substrates further extended the operable temperature up to 300°C [65]. The thermal stability of the Schottky contact and the ohmic alloys at the electrodes set an upper limit for the operation temperature of III-N rectifiers. High-temperature operation of PIN rectifiers was also studied [66]. It is shown that a homojunction GaN PIN rectifier grown on a FS-GaN substrate can achieve a blocking voltage of 800 V and a specific on-resistance ($R_{ON}A$) of 0.28 mΩ-cm^2 at the current density of 2.5 kA/cm^2. The on-state current drive of greater than 10 kA/cm^2 can be achieved in these GaN PIN rectifiers without device degradation. $R_{ON}A$ decreases with an increase of the current density, and the conductivity modulation in the drift layer was observed. The ambipolar lifetime was determined experimentally to be 9.6 ns at 25°C, and it monotonically increases to 22 ns at 175°C. The reverse-biased leakage current of a GaN PIN rectifier was attributed to the trap-assisted tunnelling with the presence of the deep-level traps (~0.7 eV) in III-N materials.

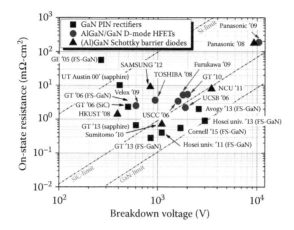

FIGURE 6.5
Comparison chart showing the breakdown voltage vs. the specific on-state resistance for different types of GaN power switches (SBD, normally-on HFETs and PIN rectifiers) grown on different substrates.

Shown in Figure 6.5 is a comparison of GaN PIN rectifiers grown on different substrates (sapphire, SiC and FS-GaN), AlGaN/GaN SBDs and normally-on AlGaN/GaN HFETs. The dashed lines in the graph represent the theoretical Baliga's figure-of-merit (BFOM) limits for Si, SiC and GaN, respectively. It is seen that III-N HFETs provide improved BFOM by at least a factor of 100 when compared with Si devices. Nevertheless, the achievable BFOM of III-N HFETs today is still approximately 10× higher than the theoretical limit of GaN. This discrepancy comes from the fact that III-N HFETs are devices in the laterally current conduction form, in which the power handling capability and the semiconductor's real-estate utilisation are not optimal. It also believed that as the voltage rating goes beyond 1 kV, III-N HFETs may not provide significant advantages over existing SiC power devices. The solutions to this issue are to develop vertically conducting devices such as HBTs or IGBTs, the device technologies that either have not been demonstrated in the case of III-N IGBTs or would require further study in the case of III-N HBTs. On the other hand, it has been demonstrated that state-of-the-art vertical GaN PIN rectifiers show a BFOM essentially reaching the theoretical limit for GaN and achieve the best power switching performance among all known semiconductor material platforms.

6.5 Radiation Effect in III-N Materials and Devices

Empirically, the radiation hardness varies inversely with the lattice constant of a semiconductor. The extent of the radiation hardness of a material can be evaluated by the threshold displacement energy. In GaN, the threshold displacement energy for gallium is ~20 eV and that for nitrogen is ~11 eV. For comparison, the threshold displacement energy of GaAs is approximately 9.8 eV. GaN devices are theoretically good choices for radiation-hard applications. Studies were conducted in III-N bulk materials and physical devices to gain insights into the radiation effect induced by various radiation sources, such as gamma, protons, heavy ions and neutrons. Depending on the intended space applications, the devices of interest will be designed to sustain functional operation under a certain level of radiation fluence and single-event effects. The radiation effects in III-N devices were characterised by comparing the phenomenological changes in the electrical and photonic performance, augmented by the trap energy analysis through deep-level transient spectroscopy (DLTS) and deep-level optical spectroscopy (DLOS) techniques.

Although the WBG nature of III-N devices has facilitated the realisation of high-temperature and high-power devices, the control of the materials synthesis and heterostructure growth is essentially a complex defect engineering endeavour. With the absence of the high-energy particle irradiation on III-N materials, as-grown semiconductors already contain various types of defects, such as vacancies, interstitials, antisites, threading dislocations and point defects. These defects are characterised by the energy states inside of the forbidden bandgap. They change the carrier transport properties through the radiative (photonic) or nonradiative (phononic) processes and the polarisation relaxation. For example, typical threading dislocation densities of III-N films grown on foreign substrates are on the order of 10^8 to 10^9 cm^{-2}. This value can be drastically reduced to 10^3 cm^{-2} on an ammonothermal GaN substrate, which is a rare find in the market due to its high cost. As such, most of the radiation studies on GaN to date have been using materials grown on foreign substrates, and the defect density in the as-grown films may be greater than those induced by the high-energy radiation. The device-level radiation effect studies were mostly performed on III-N HEMTs. Radiation effects on bipolar devices were less explored [67].

Interestingly, the radiation effect on III-N devices is not necessarily adverse. For example, silicon nitride–passivated AlGaN/GaN HEMTs were studied under a ^{60}Co irradiation of 600 Mrad for a total ionisation dose (TID) effect [68]. The results showed that the radiation possibly induced nitrogen vacancies and surface nitrogen desorption. The nitrogen vacancies were known to create shallow donor states and acceptor midgap traps. The irradiated device presented a slight negative shift of threshold voltage. However, a slight increase of the peak transconductance with no measurable changes in the mobility, sheet carrier density and contact resistance was observed. Low-dose gamma radiation could also be intentionally introduced to relax the elastic strain in AlGaN/GaN HEMTs to enhance the electron mobility. Nevertheless, as the gate length decreases, the TID effect on AlGaN/GaN HFETs becomes more pronounced.

Electron and proton radiation effects on GaN HEMTs were commonly studied. A good review of the radiation effects in III-N HEMTs can be found in [69]. Studies show that the radiation can introduce additional point defects in III-N materials. In a highly doped III-N film, the primary radiation damage defects also form a complex with the impurities, resulting in multiple defect-related energy levels [70]. Phenomenologically, AlGaN/GaN HEMTs started to show degraded device performance in terms of DC characteristics for protons with energies of 1.8–40 MeV and doses of 10^{10} to 10^{15} cm^{-2}. The postirradiation annealing does not fully restore preirradiation performance. Since the current conduction path is very close to the surface of a III-N HEMT, it was found that higher-energy protons create less displacement damage than low-energy protons. The proton irradiation, when applied properly, can be used to improve the high-voltage breakdown characteristics. When the proton is irradiated on AlGaN/GaN HEMTs grown on a Si substrate from the backside of the wafer, the off-state breakdown can be improved with no degradation in the drain current drive [71]. On the other hand, the irradiation of fast neutrons on GaN HEMTs creates a large recoil cascade, and tends to create a Fermi-level pinning around E_C −0.8 to −1.0 eV. The radiation effect is also dependent on the heterojunction designs and material growth techniques. By comparing III-N HEMTs using AlN/GaN, AlGaN/GaN and InAlN/GaN heterostructures, a binary barrier design using an AlN/GaN heterojunction shows the best carrier removal rate performance, followed by AlGaN/GaN HEMTs, and with worst performance for InAlN/GaN. The dependency of the barrier design in III-N HEMTs can be understood by the bond strength difference between AlN, AlGaN and InAlN alloys. Ga-rich or N-rich growth of III-N materials in either MOCVD or MBE showed favourable radiation hardness compared with devices grown by plasma-assisted MBE because the devices using ammonia-rich growth conditions were more susceptible to the proton radiation. It should be noted that although III-N materials present a great amount of defects and complex radiation-induced defect behaviours, the fluence for the GaN device starting to show degradation is approximately two orders of magnitude higher than that in GaAs devices for probing proton radiation up to 2 MeV [67].

6.6 Conclusions

The wide energy bandgap and high electron mobility properties for GaN and related materials enable a viable technology platform for harsh environment applications. The unique polarisation field engineering in III-N materials offers a new dimension of device designs with added complexity to exploit the desired device performance. Properly engineered III-N devices have shown superior performance for high-temperature, high-power and radiation-hard operations. The research and development of III-N technologies, however, is a relatively new field of study when compared with silicon or other compound semiconductors. Research efforts in III-N materials have led to fast commercialisation of several III-N electronic devices (e.g. III-N HEMTs, SBDs and PIN) and optoelectronic devices (e.g. solid-state lighting, UV LEDs and blue-green lasers) in the past years. However, several key transistor technologies, such as III-N HBTs and III-N IGBTs, are still in the early stages of research and development. Further development of a new III-N materials system (e.g. InAlN, AlInGaN and boron-containing III-N materials) could lead to new opportunities and expanded application space for III-N device technologies. The pursuit of high-quality epitaxy, low-cost native substrates, high-yield manufacturing, advanced packaging and thermal management continues to be a major research and development focus to validate the value proposition of the III-N materials system as an ultimate platform for electronic and optoelectronic devices for harsh environments.

References

1. O. Ambacher, B. Foutz, J. Smart, et al. Two dimensional electron gas induced by spontaneous and piezoelectric polarization in undoped and doped AlGaN/GaN heterostructures. *J. Appl. Phys.* 87 no. 1 (2000): 334–44.

2. H. Sun, A. R. Alt, H. Benedickter, et al. Ultrahigh-speed AlInN/GaN high electron mobility transistors grown on (111) high-resistivity silicon with f_T = 143GHz. *Appl. Phys. Exp.* 3 (2010): 094101.

3. S. Choi, H.-J. Kim, Z. Lochner, et al. Threshold voltage control of InAlN/GaN heterostructure field-effect transistors for depletion- and enhancement-mode operation. *Appl. Phys. Lett.* 96 (2010): 243506-1–3.

4. R. Wang, G. Li, O. Laboutin, et al. 210-GHz InAlN/GaN HEMTs with dielectric-free passivation. *IEEE Electron Dev. Lett.* 32 no. 7 (2011): 892–94.

5. S. Rajan, A. Chini, M. H. Wong, et al. N-polar GaN/AlGaN/GaN high electron mobility transistors. *J. Appl. Phys.* 102 (2007): 044501.

6. W. Lim, J.-H. Jeong, H.-B. Lee, et al. Normally-off operation of recessed-gate AlGaN/GaN HFETs for high power applications. *Electrochem. Solid-State Lett.* 14 no. 5 (2011): H205–7.

7. P. D. Ye, B. Yang, K. K. Ng, et al. GaN metal-oxide-semiconductor high-electron-mobility-transistor with atomic layer deposited Al_2O_3 as gate dielectric. *Appl. Phy. Lett.* 86 (2005): 063501.

8. S. Abermann, G. Pozzovivo, J. Kuzmik, et al. MOCVD of HfO_2 and ZrO_2 high-k gate dielectrics for InAlN/AlN/GaN MOS-HEMTs. *Semicond. Sci. Technol.* 22 (2007): 1272–75.

9. Y. Uemoto, M. Hikita, H. Ueno, et al. Gate injection transistor (GIT)—A normally-off AlGaN/GaN power transistor using conductivity modulation. *IEEE Trans. Electron Dev.* 54 no. 12 (2007): 3393.

10. Y.-L. Lee, T.-T. Kao, J. Merola, and S.-C. Shen. A remote oxygen plasma surface treatment technique for III-nitride heterojunction field-effect transistors, *IEEE Trans. Electron Dev.* 61 no. 2 (2014): 493–97.

11. Y. Cai, Y. Zhou, K. J. Chen, and K. M. Lau. High performance enhancement-mode AlGaN/GaN HEMTs using fluoride-based plasma treatment. *IEEE Electron Dev. Lett.* 26 no. 7 (2005): 435–37.

12. F. Medjdoub, M. Alomari, J.-F. Carlin, et al. Barrier-layer scaling of InAlN/GaN HEMTs. *IEEE Electron Dev. Lett.* 29 (2008): 422.

13. K. Joshin, T. Kikkawa, H. Hayashi, et al. A 174 W high-efficiency GaN HEMT power amplifier for W-CDMA base station applications. *Technical Digest, IEEE International Electron Device Meeting* (2003): 983–85.

14. Y. Pei, R. Chu, N. A. Fichtenbaum, et al. Recessed slant gate AlGaN/GaN high electron mobility transistors with 20.9 W/mm at 10 GHz. *Jpn. J. Appl. Phys.* 46 no. 45 (2007): L1087–89.

15. M. Micovic, P. Hasjimoto, M. Hu, et al. GaN double heterojunction field effect transistor for microwave and millimeter wave power applications. *Technical Digest, 2004 IEEE International Electron Device Meeting* (2004): 807–10.

16. V. Kumar, W. Lu, R. Schwindt, et al. AlGaN/GaN HEMTs on SiC with f_T of over 120 GHz. *IEEE Electron Dev. Lett.* 23 no. 8 (2002): 455–57.

17. M. Higashiwaki, T. Mimura, and T. Matsui. AlGaN/GaN heterojunction field-effect transistors on 4H-SiC substrates with current gain cutoff frequency of 190 GHz. *Appl. Phys. Express* 1 no. 2 (2008): 021103.

18. M. D. Hampson, S.-C. Shen, R. S. Schwindt, et al. Polyimide passivated AlGaN-GaN HFETs with 7.65 W/mm at 18 GHz. *IEEE Electron Dev. Lett.* 25 (2004): 238–40.

19. Y. F. Wu, A. Saxler, M. Moore, et al. 30 W/mm GaN HEMT by field plate optimization. *IEEE Electron Dev. Lett.* 25 no. 3 (2004): 1117–19.

20. Y. F. Wu, M. Moore, A. Saxler, et al. 40-W/mm double field-plated GaN HEMTs. In *2006 Device Research Conference*, State College, PA, June 26–28, 2006, pp. 151–152.

21. J. W. Chung, W. E. Hoke, E. M. Chumbes, et al. AlGaN/GaN HEMT with 300-GHz fmax. *IEEE Electron Dev. Lett.* 31 no. 3 (2010): 195.

22. S. Bouzid-Driad, H. Maher, N. Defrance, et al. AlGaN/GaN HEMTs on silicon substrate with 206-GHz fmax. *IEEE Electron Dev. Lett.* 34 no. 1 (2013): 36–38.

23. K. Shinohara, D. Regan, A. Corrion, et al. Deeply scaled self-aligned-gate GaN DH-HEMTs with ultrahigh cutoff frequency. Presented at the *International Electron Device Meeting*, Washington, DC, 2011.

24. T. Palacios, A. Chakraborty, S. Rajan, et al. High-power AlGaN/GaN HEMTs for Ka-band applications. *IEEE Electron Dev. Lett.* 26 no. 11 (2005): 781–83.

25. M. Micovic, A. Kurdoghlian, P. Hashimoto, et al. GaN HFET for W-band power applications. Presented at the International Electron Device Meeting, San Francisco, CA, 2006.

26. A. Maekawa, T. Yamamoto, E. Mitani, and S. Sano. A 500W push-pull AlGaN/GaN HEMT amplifier for L-band high power application. Presented at Proceedings of the IEEE MTT-S International Microwave Symposium Digest, San Francisco, CA, June 11–16, 2006.

27. S. Yagi, M. Shimizu, H. Okumura, et al. 1.8 kV AlGaN/GaN HEMTs with high-k/oxide/SiN MIS structure. In *Proceedings of the 19th International Symposium on Power Semiconductor Devices and ICs*, Jeju Island, Korea, May 27–30, 2007, pp. 261–64.

28. N. Ikeda, S. Kaya, J. Li, et al. High power AlGaN/GaN HFET with a high breakdown voltage over 1.8 kV on 4 inch Si substrates and the suppression of current collapse. In *Proceedings of the 20th International Symposium on Power Semiconductor Devices and ICs*, Orlando, FL, May 18–22, 2008, pp. 287–90.

29. M. Hikita, M. Yanagihara, K. Nakazawa, et al. 350V/150A AlGaN/GaN power HFET on silicon substrate with source – Via grounding (SVG) structure. *IEEE Trans. Electron Dev.* 52 no. 9 (2005): 1963–68.

30. N. Tipieneni, A. Koudymov, V. Adivarahan, et al. The 1.6 kV AlGaN/GaN HEMTs. *IEEE Electron Dev. Lett.* 27 no. 9 (2006): 716–18.

31. Y. Dora, A. Chakraborty, L. McCarthy, et al. High-breakdown voltage achieved on AlGaN/GaN HEMTs with integrated slant field-plate. *IEEE Electron Dev. Lett.* 27 no. 9 (2006): 713–15.

32. Y. Choi, M. Pophristic, B. Peres, et al. Fabrication and characterization of high-breakdown voltage AlGaN/GaN heterojunction field effect transistors on sapphire substrates. *J. Vac. Sci. Technol. B* 24 no. 6 (2006): 2601–5.

33. Y. Dora, A. Charkraborthy, L. McCarthy, et al. High breakdown voltage AlGaN/GaN HEMTs using trench gates. In *2006 Device Research Conference*, State College, PA, June 26–28, 2006, pp. 161–62.

34. C. S. Suh, Y. Dora, N. Fichtenbaum, et al. High breakdown enhancement mode AlGaN/GaN HEMTs with integrated slant field-plate. *Technical Digest, IEEE International Electron Device Meeting* (2006): 911–13.

35. Y. Uemoto, D. Shibata, M. Yanagihara, et al. 8300 V blocking voltage AlGaN/GaN power HFET with thick poly-AlN passivation. *Technical Digest, IEEE International Electron Device Meetings* (2007): 861–64.

36. W. Huang, T. Chow, Y. Niiyama, et al. Lateral implanted RESURF GaN MOSFETs with BV up to 2.5 kV. In *Proceedings of the 20th International Symposium on Power Semiconductor Devices and ICs*, Orlando, FL, May 18–22, 2008, pp. 291–94.

37. S. Chowdhury, B. Swenson, and U. Mishra. Enhancement and depletion mode AlGaN/GaN CAVET with Mg-ion-implanted GaN as current blocking layer. *IEEE Electron Dev. Lett.* 29 no. 6 (2008): 543–45.

38. M. Kanechika, M. Sugimoto, N. Soejima, et al. A vertical insulated gate AlGaN/GaN heterojunction field effect transistor. *Jpn. J. Appl. Phys.* 46 no. 21 (2007): L503–5.

39. M. Kodama, M. Sugimoto, E. Hayashi, et al. GaN-based trench gate metal oxide semiconductor field-effect transistor fabricated with novel wet etching. *Appl. Phys. Express* 1 (2008): 021104-6.

40. Y. Lee. PhD dissertation, Development of III-Nitride transistors: heterojunction bipolar transistors and fieldeffect transistors. Georgia Institute of Technology, Atlanta, May 2015.

41. Y.-C. Lee, C.-Y. Wang, T. J. Kao, and S.-C. Shen. Threshold voltage control of recessed-gate III-N HFETs using an electrode-less wet etching technique. Presented at *CSMANTECH Conference*, Boston, April 23–26, 2012.

42. D. Reusch and J. Glaser. *DC-DC Converter Handbook*. Glendale, AZ: EPC Corp., 2015.

43. T. Makimoto, Y. Yamauchi, and K. Kumakura. High-power characteristics of GaN/InGaN double heterojunction bipolar transistors. *Appl. Phys. Lett.* 84 (2004): 1964–66.

44. T. Makimoto, K. Kumakura, and N. Kobayashi. High current gain (>2000) of GaN/InGaN double heterojunction bipolar transistors using base regrowth of p-InGaN. *Appl. Phys. Lett.* 83 (2003): 1035–37.

45. K. Kumakura and T. Makimoto. High performance pnp AlGaN/GaN heterojunction bipolar transistors on GaN substrates. *Appl. Phys. Lett.* 92 (2008): 153509.

46. T. Makimoto, T. Kido, K. Kumakura, et al. Influence of lattice constants of GaN and InGaN on *npn*-type GaN/InGaN heterojunction bipolar transistors. *Jpn. J. Appl. Phys.* 45 no. 4B (2006): 3395–97.

47. T. Chung, J. Limb, J. Ryou, et al. Growth of InGaN HBT by MOCVD. *J. Electr. Mater.* 35 no. 4 (2006): 695–700.

48. R. D. Dupuis, T. Chung, D. Keogh, et al. High current grain graded GaN/InGaN heterojunction bipolar transistors grown on sapphire and SiC substrates by metalorganic chemical vapor deposition. *J. Cryst. Growth* 298 (2007): 852–56.

49. S.-C. Shen, Y.-C. Lee, H.-J. Kim, et al. Surface leakage in GaN/InGaN double heterojunction bipolar transistors. *IEEE Electron Dev. Lett.* 30 no. 11 (2009): 1119–21.

50. Y. Zhang, Y. Lee, Z. Lochner, et al. GaN/InGaN double heterojunction bipolar transistors on sapphire substrates with current gain > 100, J_C > 7.2 kA/cm^2, and power density > 240 kW/cm^2. Presented at the International Workshop on Nitride Semiconductors, Tampa, FL, September 19–24, 2010.

51. S.-C. Shen, R. D. Dupuis, Y.-C. Lee, et al. GaN/InGaN heterojunction bipolar transistors with f_T > 5 GHz. *IEEE Electron Dev. Lett.* 32 no. 8 (2011): 1065–67.

52. S.-C. Shen, R. D. Dupuis, A. Lochner, et al. Working toward high-power GaN/InGaN heterojunction bipolar transistors. *Semicond. Sci. Technol.* 28 no. 7 (2013): 074025–33.

53. Y. Zhang, Y.-C. Lee, Z. Lochner, et al. GaN/InGaN heterojunction bipolar transistors with collector current density > 20 kA/cm^2. *Technical Digest, 2011 CSMANTECH Conference* (2011): 201–4.

54. Y. Lee, Y. Zhang, Z. Lochner, et al. GaN/InGaN heterojunction bipolar transistors with ultra-high d.c. power density > 3 MW/cm^2. *Phys. Stat. Solidi A* 209 no. 3 (2012): 497–500.

55. D. Keogh, P. Asbeck, T. Chung, et al. High current gain InGaN/GaN HBT with 300 °C operating temperature. *Electron. Lett.* 42 (2006): 661–63.

56. K. Kumakura and T. Makimoto. High-temperature characteristics up to 590°C of a pnp AlGaN/GaN heterojunction bipolar transistor. *Appl. Phys. Lett.* 94 (2009): 103502.

57. T. G. Zhu, D. J. Lambert, B. S. Shelton, et al. High-voltage GaN pin vertical rectifiers with 2 μm thick i-layer. *Electron. Lett.* 36 no. 23 (2000): 1971–72.

58. J. B. Limb, D. Yoo, J. H. Ryou, et al. Low on-resistance GaN pin rectifiers grown on 6H-SiC substrates. *Electron. Lett.* 43 no. 6 (2007): 67–68.

59. Y. Saitoh, K. Sumiyoshi, M. Okada, et al. Extremely low on-resistance and high breakdown voltage observed in vertical GaN Schottky barrier diodes with high-mobility drift layers on low dislocation-density GaN substrates. *Appl. Phys. Exp.* 3 no. 8 (2010): 081001-3.

60. A. P. Zhang, J. W. Johnson, B. Luo, et al. Vertical and lateral GaN rectifiers on free-standing GaN substrates. *Appl. Phys. Lett.* 79 no. 10 (2001): 1555.

61. I. C. Kizilyalli, A. P. Edwards, H. Nie, and D. P. Bour. High voltage vertical GaN p-n diodes with avalanche capability. *IEEE Trans. Electron Devices* 60 no. 10 (2013): 3067–3070.

62. Y. Hatakeyama, K. Nomoto, N. Kaneda, et al. Over 3.0 GW/cm^2 figure-of-merit GaN p-n junction diodes on free-standing GaN substrates. *IEEE Electron Dev. Lett.* 32 no. 12 (2011): 1674–76.
63. Y. Hatakeyama, K. Nomoto, A. Terano, et al. High-breakdown-voltage and low specific on resistance GaN *p–n* junction diodes on free-standing GaN substrates fabricated through low-damage field-plate process. *Jpn. J. Appl. Phys.* 52 (2013): 028007.
64. W. Lian, Y. Lin, J. Yang, et al. AlGaN/GaN Schottky barrier diodes on silicon substrates with selective silicon diffusion for low onset voltage and high reverse blocking. *IEEE Electron Dev. Lett.* 34 (2013): 981.
65. Z. Xiaoling, L. Fei, L. Changzhi, et al. High temperature characteristics of Al$_x$Ga$_{1-x}$N/GaN Schottky diodes. *J. Semicond.* 30 (2009): 034001-1.
66. T. Kao, J. Kim, Y. Lee, et al. Temperature-dependent characteristics of GaN homojunction rectifiers. *IEEE Trans. Electron Dev.* 62 no. 8 (2015): 2679.
67. A. Ionascut-Nedelcescu, C. Carlone, A. Houdayer, et al. Radiation hardness of gallium nitride. *IEEE Trans. Nucl. Sci.* 49 no. 6 (2002): 2733–38.
68. O. Aktas, A. Kuliev, V. Kumar, et al. ^{60}Co gamma radiation effects on DC, RF, and pulsed I-V characteristics of AlGaN/GaN HEMTs. *Solid-State Electron.* 48 no. 3 (2004): 471.
69. S. Pearton, Y. Hwang, and F. Ren. Radiation effects in GaN-based high electron mobility transistors. *JOM* 67 no. 7 (2015): 1601–11.
70. A. Sasikuma, A. R. Arehard, S. Kaun, et al. Defects in GaN based transistors. *Proc. SPIE* 8986 (2014): 89861C.
71. Y. Hwand, S. Li, Y. Hsieh, et al. Effect of proton irradiation on AlGaN/GaN high electron mobility transistor off-state drain breakdown voltage. *Appl. Phys. Lett.* 104 (2014): 082106.
72. Y. Zhang. PhD dissertation, Development of III-Nitride bipolar devices: avalanche photodiodes, laser diodes, and double-heterojunction bipolar transistors. Georgia Institute of Technology, 2011.
73. Edgar, J. H. *Properties of Group III Nitrides*. London: INSPEC, 1994.
74. F. Bernardini, V. Fiorentini, and D. Vanderbilt. Accurate calculation of polarization-related quantities in semiconductors. *Phys. Rev. B* 63 no. 19 (2001): 193201.
75. A. F. Wright. Elastic properties of zinc-blende and wurtzite AlN, GaN, and InN. *J. Appl. Phys.* 82 (1978): 2833.

Section III

Packaging and System Design

7

Packaging for Systems in Harsh Environments

Marc Christopher Wurz and Sebastian Bengsch

CONTENTS

ABSTRACT   The request for packaging technologies which enable electronic devices to operate under harsh environments is always increasing. The main challenge is the encapsulation of electronic circuits to save them from external thermal influences and to resist fast thermal gradients. Other aspects of harsh environments are radioactive, chemical, electromagnetic and high-pressure surroundings. For all of these special influences, different embodiments for the packaging must be chosen. For example, for industrial applications ceramic is often used to ensure a long-term stability of about 10 years. This chapter gives an overview of the different techniques for the fabrication of packages for operation under harsh environments.

7.1 Introduction

The field of industrial applications which operate in a harsh environment has increased significantly. Against this background, this chapter gives an overview of the possible materials which allow operation under different harsh environment conditions, such as thermal, chemical, electromagnetic and high-pressure loads. For each application, different materials can be used, such as ceramics, metals or silicon. Every material needs a specific fabrication technology. which is described in the following.

7.2 Definition of Harsh Environments

Harsh environments are generally known as environments that have environmental impacts on electronic devices under high or low pressures or extremely high or low temperatures. Other harsh environments are radioactive, chemical and electromagnetic surroundings. Harsh environments are known in the fields of oil and gas exploration and production; undersea cabling; and industrial, medical and aerospace technology. The space environment is a very good example of a combination of challenges which have to be met. Space has very high temperature gradients, very low pressures and strong electromagnetic fields which influence electronic devices used in satellites or spacecrafts. In addition to the environmental influences, the electronic devices used have to endure a much longer lifespan than consumer electronics, with no maintenance procedures. Regarding the production of electronics, space agencies have strong requirements for the reliability of the applied technology. Other known harsh environments occur during mechanical processing, for example, in the application of sensors in drills or grinders where high temperatures, pressures and chemical atmospheres are possible. Under the influence of lubricant solvents, high-temperature gradients and fast cooling of the material result in stress damage. Regarding the oil and gas industry, very high pressures are becoming a challenging factor, increasing with the depth of the drill, besides the temperatures, chemical surroundings and radioactive contamination. Difficulties increase in underwater drilling where pressures of up to 1000 atm appear. For industrial standardisation, protection classes for electrical and electronic devices are categorised by the International Protection Codes (IP or IK). Aerospace and military applications have their own regulations in addition to the IP and IK classes. In aerospace research, important parameters have to be analysed directly inside the jet turbine engine, where temperatures up to 2200°C occur. These measurements are highly important for analysing and increasing the efficiency of jet turbine engines. Flow velocities of 150 m/s and pressures of up to 43 atm are found in an Airbus A380 turbine engine and are common values in commercial aircraft technology. Only a few materials are capable of handling temperatures above 2000°C, and interconnection between devices becomes a further challenge at this point. All these factors limit the choice of material, manufacturing and bonding technology. Therefore, the development of new techniques, materials and adhesives in the field of packaging technologies for harsh environments is the focus of much research. General requirements of housing material are [1,2]

1. High electrical resistance
2. Low relative dielectric constant
3. Low loss factor
4. High mechanical stability
5. High thermal conductivity
6. High-temperature-change stability
7. No radioactive components
8. Good metallisation
9. Dimensional accuracy

7.3 Possibilities for Packages

Generally known housing materials for harsh environments are ceramics, metals and silicon; each material has its own benefits regarding atmosphere, pressure stability and temperature influences. Investigations of all silicon-based substrates and housings are the focus of research. General problems that occur while combining different materials, such as metals and ceramics, are different thermic expansion coefficients and the joining technique between two objects. When it comes to metal-based components, machined housings are common state of the art for highly precise and low-number lots. One difficulty occurring in metal housings is contact outside of the housing. A solution for outside contact may be wireless signal transportation at high-temperature localisation. As a basis for high-temperature environment applications, SiC substrates are used. The main advantages of SiC are a high-temperature stability of up to 2500°C, resistance to chemical attack (acids and hot gases) and the ability to handle 1 GPa of yield stress and allow elastic deformation in the small-deflection regime due to pressure. Recent research shows a possible thermoshock capability of 1000°C in 3 s. Simple bonding technologies like soldering can be performed with SiC, and solder with gold and tin is the main focus of research. Besides mechanical or chemical structuring of SiC to manufacture housing components for electronic devices,

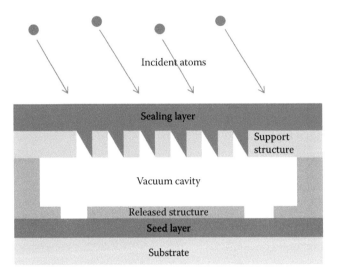

FIGURE 7.1
Sputter deposition encapsulation.

sputter deposition is a new approach in encapsulation and offers a thin-film technology-based production technology for harsh environment packaging, which is shown in Figure 7.1. The atoms impact on the substrate surface. This closes the openings in the scaffold layer, so the packaging is fabricated under a high vacuum, which results in the encapsulation of this vacuum in the cavity.

In the case of thick-film technology, ceramic substrates like Al_2O_3 are carriers of surface- mounted and unhoused devices. Thick-film technology offers the advantage of resistors directly integrated into the substrate level. The main tasks of the ceramic substrates are being a carrier of connectors and being a carrier of Surface Mount Device (SMD), discretely or directly integrated, and therefore manufactured by printing processes. Additionally, ceramic substrates are the insulating factor between electric components. The main advantages of ceramic substrates are their good planarity of less than 0.4%, adapted thermic expansion coefficient and good thermic conductivity of more than 20 W/mK. Excellent temperature resistance up to 1000°C and a high specific resistance of more than 10^{-7} Ω*m, as well as a high flexural strength of 200 N/mm^2, are only a few benefits.

Usage at temperatures higher than 1000°C is possible. The relative permittivity is 9–10 at 100 MHz, and the loss factor is 10^{-4}. The melting point is at 2072°C; therefore, the application temperature should be below 1900°C. The mechanical stability depends on the clarity of the aluminium oxide and its atomic lattice. It is generally known that the higher the clarity the substrate offers, the better its mechanical stability. On the other hand, the production process complexity rises with the clarity of the substrate. Aluminium oxides

TABLE 7.1

Properties of Different Substrate Materials

	AL_2O_3	BeO	AIN	EES
Maximum process temperature (°C)	1500	1800	1400	550–650
Thermal expansion coefficient (10^{-7}/K)	75	85	34	90
Thermal conductivity (w/mK)	20	230	150	60–80
Flexural strength (N/mm²)	320	170	300	—
Surface roughness (μm)	0.5	0.5	1–5	—
Specific electric resistance (Ω at 20°C)	10^{14}	$>10^{15}$	10^{13}	$>10^{14}$
Dielectricity	9.5	7.0	10.0	6–8
Loss factor at 1 MHz	$3*10^{-4}$	$2*10^{-4}$	$2*10^{-3}$	$3*10^{-3}$ to $6*10^{-3}$
Cost factor	1	50	25	2

offer great properties in terms of surface quality and tribological and abrasive behaviour. A big disadvantage of aluminium oxides is their dissolvable behaviour in strong acids and strong leach. Therefore, aluminium oxides are not suitable for harsh atmospheres and chemically influenced environments.

Besides silicon, SiC, aluminium oxides and metals are an important part of harsh environment packaging. There are applications and metals that combine resistance with harsh environmental impacts. Research and industrial application show the use of high-strength aluminium or titanium as housing materials. The housing material is mechanically machined by computer numerical control (CNC) milling or rotation. Big disadvantages are the high costs and time periods for big lots. On the other hand, magnetic influences can be limited by special alloys like NiFe, which depends on the permeability of the material and is not provided by any ceramic or silicon-based material, and therefore is only reserved to metals. The different mechanical properties and thermic expansion coefficients of different materials are shown in Table 7.1 and Figure 7.2 [3]. Figure 7.3 shows the ceramic packaging technology used for all ceramic-based substrates and chips which have also been CNC machined beforehand. Table 7.2 shows the comparison between different housing materials.

Bonding technologies of harsh environment packaging are the main focus of research because of the gained system stability which is provided by the bonding technology to seal the housing. Table 7.3 shows the comparison of mechanical bonding technologies. The main problems of the current bonding technologies are the low-temperature resistances and therefore the application in high-temperature harsh environments. An interesting bonding technology is silicon fusion bonding. Bonding temperatures are between 700°C and 1000°C. The application temperature of a bonded housing must be below 700°C. Difficulties with this bonding technology are the conditions of

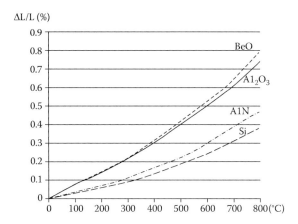

FIGURE 7.2

Comparison of thermic expansion coefficients of different ceramics. (From Fhm, Mikroelektronik/Technologie: Dickschicht-Hybridtechnik, 2015.)

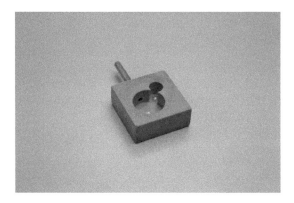

FIGURE 7.3

Ceramic housing.

TABLE 7.2

Comparison of Housing Materials. Properties Are Indicated by [Minus Sign] = Negative [Plus Sign] = Positive and 0 = Neutral

	Silicon	Gallium Arsenide	Aluminium Nitride	Silicon Carbide
Electrical	Good below 150°C, –	Good to 350°C, o	Difficult to make circuits, o	Good through 600°C, +
Mechanical	Softens at high temperatures,–	Weaker than silicon, –	Stable past 700°C, +	Stable past 700°C, +
Chemical	Can be etched,–	Several wet etchantsk –	Can be etched, –	Robust, +

Source: Hjort, K., et al., *Journal of Micromechanics and Microengineering,* 4(1), 1–13, 1994; Slack, G. A., and Bartram, S. F., *Journal of Applied Physics,* 46(1), 89–98, 1975; Mehregany, M., *Proceedings of the IEEE,* 86(8), 1594–1610, 1998.

TABLE 7.3

Comparison of Bonding Technologies

Method	Materials	Intermediate Layers	Temperature (°C)	Surface Preparation	Selective Bonding
Low-temperature direct bonding	Si-Si, SiO_2-SiO_2		200–400	Plasma treatment, wet surface activation (dip)	
Eutectic bonding		Au, Al	379, 580	Sputtering, electroplating	Lift-off, etching
Welding	Si-Si	Au, Pb-Sn	300	Evaporation, sputtering	Lift-off, etching
Adhesive bonding	Si-Si, Si-glass, SiO_2-SiO_2, Si_3N_4-Si_3N_4	Adhesives, photoresist	RT-200	Spincoating	Lithography
Anodic bonding	Glass-Si, Si-Si, Si-metal/glass	Pyrex; sputtered Al; W, Ti, Cr	>250, >300, 300–500	Voltage, 50–100 V	Lithography, etching, lift-off
Silicon fusion bonding	Si-Si, SiO_2-SiO_2		700–1000	Standard cleaning	Lithography, etching
Low-temperature glass bonding	Si-Si, SiO_2-SiO_2	Na_2O-SiO_2 and other sol-gel materials, boron glass	200–400, >450	Spincoating CVD, implantation	Lithography, etching

the surface, which have to be extremely precise and flat (roughness between 1 and 4 nm). The bonding steps are

- Phase 1: Below 300°C, van der Waals bonding occurs. OH molecules located on the surface bond to oxygen.
- Phase 2: Above 300°C, Si-O-Si bonds develop. Both wafers grow together. Equalisation of roughness through elastic deformation occurs.
- Phase 3: At higher temperature, viscous yielding of the oxide occurs. A full bond force appears.

7.4 Fabrication Process for Packages Depending on the Environment

7.4.1 Environmental Conditions for Material Choices

Housing technologies have to be adapted to their application surroundings and withstand environmental impacts. Certain materials can only resist specific influences, such as chemical atmosphere, temperature and pressure. Generally, plastics are able to withstand chemical surroundings, like inorganic acid and leach, whereas solvents are a big problem for most plastic components; plastics are also only applicable in low-temperature environments. Metals are capable of withstanding higher temperatures in comparison with plastic, but common metals insufficiently withstand inorganic acids. Ceramics as a material choice offer excellent temperature stability and chemical resistance, but they are difficult to manufacture and mechanically process. Therefore, the housing technology used has to be resistant to its application surroundings and has to be chosen wisely.

7.4.2 High-Precision CNC Machining

High-precision CNC machining is a well-known but expensive technology to manufacture metal-based housings for harsh environmental applications. The chosen metal is CNC rotated or milled into the desired shape and assembled with the electronic device. Modern CNC milling machines have a precision of 2.5 microns and can handle various materials, for example, titanium, cobalt–chromium or composite materials. The use of CNC-machined materials is a common process for heat, pressure and flow sensors for turbine engine analysis. Therefore, the application sensor must resist high temperatures and be highly precise without any damage occurring to the tools, machine or material.

7.4.3 Ceramic Al_2O_3 Housing

The ceramic sintering process for housing technologies is comparable to the manufacturing process of ceramic substrates of Al_2O_3, which is shown

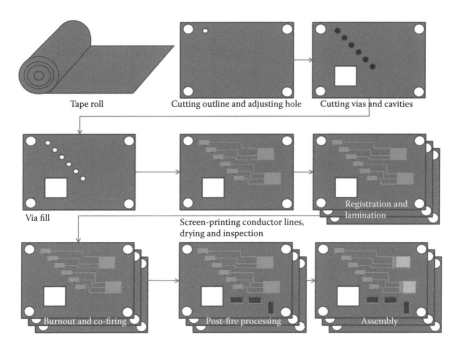

FIGURE 7.4
Manufacturing process of ceramic substrates.

in Figure 7.4. The ceramic housing is produced from a ceramic material of a tape roll, cut into shape and stacked. Following the preparation, the ceramic housing is fire processed and assembled with the electronic devices. Afterwards, the housing is sealed using different sealing options. The difficulties that occur are contact outside of the housing, as well as bonding and sealing of the housing itself. Ceramics are also limited in their flexibility, and mechanic influences like vibration can cause damage to the housing. This can result in the loss of the electronic device. A combination of flexible bonding technologies and ceramic materials can be a solution for this inflexibility. Compound materials are also the focus of research. Al_2O_3 is a well-known ceramic material, and it already offers standardised manufacturing processes, which are produced in big quantities. Physical and chemical stability vary depending on the purity of the Al_2O_3. The higher the purity, the better the properties. Al_2O_3 ceramics fulfil the general requirements of housing materials. The slip of the tape roll consists of the following ingredients [7]:

1. Ceramic powder (ca. 50%)
2. Solvent: Water
3. Plasticiser: Glycerin, glycol
4. Deflocculant: NaCl aryl sulphonacid
5. Surface-active agent: Octylphenoxyethanol

The manufacturing process of Al_2O_3 housing consists of the following steps [9]:

1. Production of the slip for a tape roll
2. Manufacturing of the ceramic tape roll via the doctor blade process
3. Stripping of the tape roll from a steel band
4. Cutting and pinching via CO_2 laser
5. Screen printing of the tape roll with paste (paste consists of 90% metal powder + 10% glass powder)
6. Laminating of the tape roll
7. Post–fire processing under hydrogen
8. Further assembly

7.4.4 Silicon Wafer Housing

Silicon wafer housing is a method to minimise the thermal stress and differences in the thermal expansion coefficient. Silicon substrates are also a well-known technology and easy-to-handle material. The substrate is structured by chemical wet etching, and the behaviour of isotropic etching is used to form a valley profile (Figure 7.5). The lowered profile is the top of the housing, which is mounted on a common silicon wafer by wafer bonding processes under vacuum. Further advantages include the standardised production technologies for silicon substrate–based materials. The bonding technologies must fulfil several factors which are provided by only a few adhesives. Adhesives for harsh environments are the focus of research [8].

7.4.5 Ceramic SiC Housing

As already mentioned in Chapter 2, the highly interesting material SiC is the focus of research, as barely any substrate has the excellent harsh environment performance of silicon carbide. SiC is produced in different ways, depending

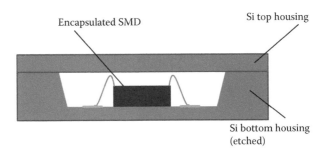

FIGURE 7.5
Schematic illustration of all-silicon packaging scheme.

on its application. With the technology of chemical vapour deposition (CVD), thin layers of SiC can be applied on different substrates. For example, there are mirrors in space telescopes which are coated with SiC because of its very small thermic expansion coefficient. For housing, different technologies have to be used. In general, there can be differences between two forms of SiC. SiC is a ceramic material and can be open porous or dense.

Open porous examples are silicate-bonded silicon carbide, recrystallised silicon carbide (RSIC) and nitride- or oxynitride-bonded silicon carbide (NSIC).

Dense SiC examples are reaction-bonded silicon carbide (RBSIC), silicon-infiltrated silicon carbide (SISIC), sintered silicon carbide (SSIC), hot (iso-static) pressed silicon carbide (HPSIC [HIPSIC]) and liquid-phase sintered silicon carbide (LPSIC).

SiC is processable as a sintered ceramic, and manufacturing processes start from pure SiC powder with nonoxide sintering aids. The material is sintered in an inert atmosphere at temperatures up to 2000°C or higher. The melting point at 2500°C indicates the high percentage of covalent bonding of the ceramics and shows the high thermic stability of this material. The different SiC ceramics vary in their production processes and application.

Known properties of SiC ceramics are [7]

- Very high hardness
- Corrosion resistance, even at high temperatures
- High resistance to wear
- High strength, even at high temperatures
- Resistance to oxidation, even at very high temperatures
- Good thermal-shock resistance
- Low thermal expansion
- Very high thermal conductivity
- Good tribological properties
- Semiconductivity

7.4.6 Design Examples and New Process Technologies for Harsh Environmental Packaging

Harsh environmental packaging is already a common field in space and aircraft technology. One piece, the CNC-machined metal housing, is common state of the art in aircraft technology. In satellite technology, all-ceramic electronics, with ceramic housings, offer the required benefits for the space environment. Companies like Ametek or Siennatek specialise in harsh environment packaging.

New process technologies include ideas such as directly sputtered housings as a thin-film layer on the substrate or electronic device, as mentioned in Chapter 2. New approaches are in the area of functional ceramics like SiC,

(a)

(b)

FIGURE 7.6
(a) Indium–glass bonding. (b) Indium–ceramic bonding.

which offer various manufacturing processes which can individualise the benefits of the material in terms of its application. Especially at the Institute for Micro Production Technology (IMPT), bonding technologies have been the recent focus of research. Therefore, indium sealing on stainless steel housings might offer excellent opportunities for harsh environment sealing and bonding technology. Figure 7.6a and b shows glass and ceramic vacuum bonds in combination with indium on stainless steel substrates. In the case of SiO_2 and Al_2O_3, no leakage of nitrogen could be detected; the partial pressure of He was below 5×10^{-9} Torr.

7.5 Conclusion

Modern packaging technology has to face harsh environmental impacts in various fields. The big challenges appear in the oil and gas industry, aerospace and military or industrial applications. Besides high temperatures and gradients, chemical atmosphere and electromagnetic and radioactive surroundings can damage electronic devices. Special materials have to be used to resist these influences; there is no material which offers a perfect solution to all impacts at once, but a smart combination of properties can withstand the worst environments known. Space technology also has to face the fact that the maintenance of the electronics is not possible, and with even farther distances to reach, the electronic devices will have a highly increased lifespan in comparison with consumer electronics, which will only offer a lifespan of a couple of years. A big problem which is the focus of research is the interconnection and bonding technology between the electronic parts of the device and its housing. Besides the development of new brazing technologies for bonding issues, research shows that wireless communication between different devices can be a solution for interconnection problems as well. A material which is the focus of research is SiC; it provides excellent properties and benefits, making it the number one harsh environment packaging material. High temperature stability and surface quality are only a couple of the advantages of this highly interesting material.

References

1. Amtek Interconnects website. Accessed August 3rd, 2016. http://www.ametekinterconnect.com.
2. Sporian Microsystems, Inc. website. Accessed August 3rd, 2016. www.sporian.com.
3. Fhm. Mikroelektronik/Technologie: Dickschicht-Hybridtechnik, 2015.
4. K. Hjort, J. Söderkvist, J. Å. Schweitz. Gallium arsenide as a mechanical material. *Journal of Micromechanics and Microengineering*, 4(1), 1–13, 1994.
5. G. A. Slack, S. F. Bartram. Thermal expansion of some diamond like crystals. *Journal of Applied Physics*, 46(1), 89–98, 1975.
6. M. Mehregany. Silicon carbide MEMS for harsh environments. *Proceedings of the IEEE*, 86(8), 1594–1610, 1998.
7. Think Ceramics Website. Accessed August 3rd, 2016. www.keramverband.de.
8. C. Jia, J. Bardong, C. Gruber, et al. Wafer-level packaging for harsh environment application. Presented at the 4th IEEE International Workshop on Low Temperature Bonding for 3D Integration, Tokyo, July 15–16, 2014.

8

Corrosion Resistance of Lead-Free Solders under Environmental Stress

Suhana Mohd Said and Nor Ilyana Muhd Nordin

CONTENTS

ABSTRACT This chapter discusses the mechanism of corrosion of lead-free solders in harsh environments, such as marine environments or acid rain. Lead-free solders are susceptible to galvanic corrosion in the presence of humidity, resulting from the presence of dissimilar elements in their microstructure. The mechanism of galvanic corrosion is explained in terms of electrochemical migration, with potentiodynamic polarisation being the main mode of investigating the mechanism of corrosion. A few case studies on the corrosion behaviour of different lead-free solders are illustrated. In addition, some experimental methods to investigate corrosion behaviour are elaborated. Finally, a discussion on industrial standards in microelectronics packaging subjected to harsh environmental conditions is presented, in terms of performance, reliability and functionality.

8.1 Lead-Free Solders

With the enforcement of the Restriction of Hazardous Substances (RoHS) in 2006 within the electronics industry, lead-free solders have steadily gained market share in electronics packaging applications. Prior to this, Sn-Pb solders had been the industrial solder of choice. The soldering technology involving Sn-Pb alloys had been developed and improved over the years, which provided many advantages: high ductility, high wettability, ease of handling and low processing temperatures (Abtew and Selvaduray, 2000). Figure 8.1 shows the trend of lead-free solders replacing the usage of tin–lead solders in recent years.

Various binary, ternary and quaternary lead-free solder formulations based primarily on tin alloys have been the subject of intense research. For example, the ternary Sn–Ag–Cu (SAC) alloy is a popular formulation, given its thermal cyclic and drop impact reliability. In addition, doping with minor elements such as Ni, In, Bi, Mn, Sb, Ga, Fe and Al has served to improve the solder's performance. For example, the SAC-Ni and Sn-Ag-Bi (Kotadia et al., 2014) formulations have been shown to improve wetting properties, while SAC-Fe formulations have been shown to limit diffusion (Liu et al., 2014).

Generally, it has been the case that each of these formulations requires some trade-off in terms of performance and reliability issues. For example, Sn-58Bi, which was a leading candidate to replace the Sn-Pb formulation, demonstrates degradation under the thermal ageing process. In this formulation's microstructure, the interface between Cu and Cu-Sn intermetallics undergoes Bi segregation, which then ultimately leads to embrittlement.

The incorporation of lead-free solder alloys on board-level joints poses a different set of reliability issues. Figure 8.2 shows the application of a bulk solder in ball form attached to a substrate to form an interconnect. The reliability of a board-level joint is typically determined by the properties of the interfacial layer between the solder and substrate. When the solder joints are in service, there is a complex interfacial evolution between the solder bulk and substrate, such as the growth of intermetallic compound (IMC) layer thickness, the coarsening of the solder grain and the mechanical stress in the microstructure. This will greatly influence the fatigue resistance and failure mode of solder joints. The microstructural characteristics of the IMCs at the interfacial layer play a key role in crack propagation. The IMCs' generally brittle and hard, initiated cracks are likely to propagate until total fracture.

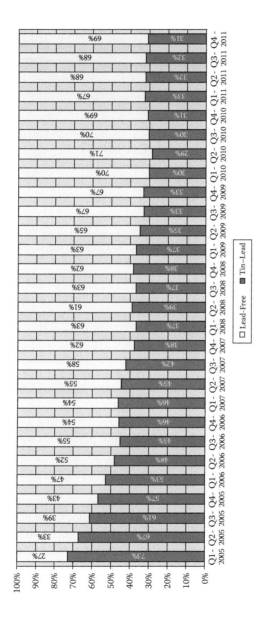

FIGURE 8.1
Tin-lead vs. lead-free solder consumption trend. (Reproduced from IPC, Electronics Industries Market Data Update Report, 2012. With copyright permission. Published by www.ipc.org)

FIGURE 8.2
Solder-to-substrate interconnect.

A variety of environmental stress factors, such as temperature and humidity, shock and vibration, and thermal fluctuations, may also influence the occurrence of solder joint failure. The most commonly observed failure modes are fatigue and overload. Overload failure occurs when the stress in the solder joint caused by the imposed stress factors is greater than the inherent strength of the solder alloy. Mechanical and electrical failures are quite commonly the cause of many system failures. In this chapter, the corrosion of lead-free solders subjected to environmental stress, particularly corrosive environments, is discussed in detail.

8.2 Corrosion of Lead-Free Solders

In many situations, electronic components are subjected to corrosive environments. In particular, automobile, marine and aeronautical applications expose these devices to corrosion media such as moisture, sulphur compounds and pollutants. Marine applications are particularly susceptible to corrosion due to the high percentage of chlorine present in seawater.

8.2.1 Types of Corrosion

In aqueous or ambient environments, most metals will spontaneously oxidise to form surface oxides. In the case of solder alloys, the presence of surface oxide adversely affects a solder joint in terms of its solderability and wettability. In electronics packaging, flux material is used to reduce surface oxides and protect them from oxidation during reflow.

The different types of corrosion include

- Pore and creep corrosion of base metals plated with a noble metal.
- Fretting corrosion. This arises when two surfaces experience both wear and corrosion damage due to mechanical loading and repetitive surface motion (e.g. vibration).

- Stress corrosion in the presence of aggressive contaminants. One example is from residual chemicals which are a by-product of the manufacturing process, many of which contain chlorides.

- Corrosion due to a combination of pollutants and moisture. Pollutants include chlorides, sulphates, sodium and calcium. The presence of humidity may induce a moisture film to coat the substrate and hence increase the corrosion rate. For example, Tompkins (1973) reported that tin nitrate was produced in air by 30%–35% humidity containing 10 ppm NO_2.

- Galvanic corrosion due to the combination of metals with dissimilar potentials. This process is electrochemically driven where it is in the presence of an electrolyte, and one metal corrodes in preference over the other.

- Electrolytic corrosion from applied potentials in electronics devices. This is an accelerated corrosion process which occurs between two metal contacts connected by an electrolyte. It is also subjected to a current flow from an electromotive force (EMF). Similarly, one metal is corroded in preference over the other.

The impact of these corrosions is particularly detrimental in harsh environmental conditions, as these corroded regions act as initiation sites for crack propagation. This will eventually lead to device failure.

Lead-free solder alloys are particularly susceptible to galvanic corrosion. This is due to their unique microstructure where the β-Sn matrix is interspersed with IMCs of varying potentials. The less noble metal tends to be Sn, which acts as the anode and is thus eroded, leaving behind IMCs such as Ag_3Sn.

8.2.2 Galvanic Corrosion

Seawater contains large amounts of salts, including aggressive Cl–, where it provides a highly corrosive environment. Similarly corrosive is acid rain where it is in the form of sulphuric and nitric acid, a product of rainwater with adsorbed industrial gas emissions such as sulphur dioxide and nitrogen oxides.

The base material for solders such as tin is regarded as highly resistant to corrosion due to the formation of a passivation layer in the presence of humidity or an aqueous environment. However, the addition of minor alloying elements may have an adverse effect on the corrosion resistance. Galvanic corrosion arises from the connection of two dissimilar metals (and hence potentials) by an electrolyte. In the case of lead-free solders, they are more vulnerable to galvanic corrosion than lead-containing solders. This is due to the presence of IMCs containing silver, such as Ag_3Sn. Tin acts as the anode and is corroded, while deposition occurs at the IMC, which acts as the cathode. In Section 8.2.3, the mechanism of galvanic corrosion through electrochemical migration (ECM) is discussed.

8.2.3 Electrochemical Migration

Ionic migration, better known as ECM, drives the mechanism of galvanic corrosion. ECM occurs in the presence of moisture or aqueous media; that is, if the two conducting (metallic) points are connected by a water pathway, the water dissolves the metal and acts as an ion channel between the two points. The two adjacent conducting points become electrically connected to form a corrosion cell. ECM causes a metallic dendrite to form, which then bridges the points and forms an electrical short circuit under the right conditions. The shape of the connection may be classified as either a dendrite or a conductive anodic filament (CAF), depending on the geometry of the resulting deposit. The cathode is the deposition site and electrodeposition starts when metal ions travel within the electrolytes in the presence of current. There are three steps in migration: (1) the anodic reaction, (2) the cathodic reaction and (3) the interelectrode reaction.

Conditions of the ECM include the thermodynamic nature and the stability of these metal ions. The failures from ECM are typically intermittent, since it is triggered by the presence of aqueous media. Contact may also be lost once the water layer has evaporated and high currents are transported through the dendritic structure.

Electromigration of tin and its alloys is of particular interest, and there have been several studies on the corrosion of lead-free solders. Yu et al. (2006) investigated Sn-37Pb andSn-36Pb-2Ag, comparing them with the Sn-Ag and Sn–Ag–Cu systems. In the lead-containing formulations, the main migration element was Pb, while for the lead-free alloy, the main migration element was Sn. In a Sn-Pb system, pure lead was shown to be the most susceptible to ECM, and similar behaviour was shown for alloys containing up to 60% Sn. For higher percentages of tin, less or no ECM was observed.

A higher surface roughness would encourage the dissolution of tin. Previous works by Ambat et al. (2009), Johnsen et al. (2009) and Minzari et al. (2009) have also shown that the structure and resistance of the dendrites are a function of the chemistry of the electrolyte and potential bias. In particular, the dendritic structure depends on the current density and the metal ion concentration. The formed dendrites are as illustrated in Figure 8.3.

8.2.4 Potentiodynamic Studies for Corrosion Investigation

In Section 8.2.3, galvanic corrosion was described in terms of a galvanic cell, where the presence of two dissimilar metals that are connected by an electrolyte causes one metal to be eroded in preference over the other. Potentiodynamic analysis, also known as potentiodynamic polarisation, allows a qualitative method to evaluate the dissolution of the metal into the

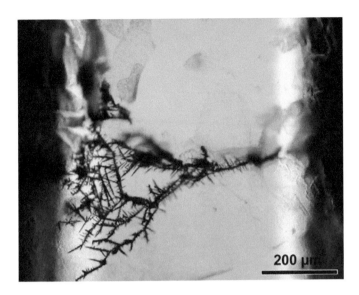

FIGURE 8.3
Dendrites after the water drop test. (Reproduced from Medgyes, B., et al., *Corrosion Science* 92: 43–47, 2015. With copyright permission. Published by Elsevier.)

electrolyte. This is a polarisation method in which the corrosion behaviour of the metal–electrolyte system is simulated in laboratory conditions. In order to identify the potential, the electrode is varied within a large potential domain at a specific rate by the application of a current through the electrolyte. Through this, information on corrosion mechanisms, rate and susceptibility of the specific materials to corrosion in designated environments can be obtained.

Corrosion normally occurs at a rate determined by an equilibrium between opposing electrochemical reactions. One reaction is anodic, in which a metal is oxidised, releasing electrons into the metal. The other is cathodic, in which a solution species (often O_2 or H^+) is reduced, removing electrons from the metal. When these two reactions are in equilibrium, the flow of electrons from each reaction is balanced, and no net electron flow (electrical current) occurs. The two reactions can take place on one metal or on two dissimilar metals (or metal sites) that are electrically connected.

Potentiodynamic measurements were carried out using a three-electrode measuring system: the Pt counterelectrode, the SAC alloy as the working electrode and the saturated calomel electrode (SCE) as the reference. Figure 8.4 gives the schematic of the electrode configuration for the potentiodynamic polarisation study.

Figure 8.5 sketches the output of the potentiodynamic polarisation experiment. The vertical axis is the electrical potential, and the horizontal axis is the logarithm of absolute current. The theoretical current for the anodic and cathodic reactions is represented as straight lines. The curved line is the total

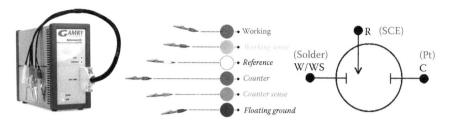

FIGURE 8.4
Setup for potentiodynamic polarisation using three electrodes by Gamry potentiostat.

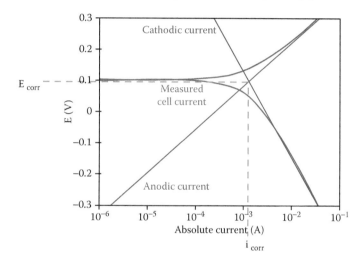

FIGURE 8.5
Potentiodynamic polarisation curve with cathodic and anodic components.

current, which is the sum of the anodic and cathodic currents. This is the current that is measured during the potential sweep by the potentiostat. The sharp point in the curve is in fact the point where the current reverses polarity as the reaction changes from anodic to cathodic, or vice versa. Typically, due to the passivity phenomenon, the current often changes by six orders of magnitude during a potentiodynamic polarisation study.

The corrosion rate, CR (mm/year), can be determined from Equation 8.1:

$$CR = 3.27 \times 10^{-3} \left(i_{corr} EW \right) / \rho \tag{8.1}$$

where EW is the equivalent weight of the corroding species in grams, and ρ is the density of the corrosive material in grams per cubic centimetre.

At the anode, dissolution of the metal M occurs as follows through oxidation of the metal:

$$M \rightarrow M^{n+} + ne^{-} \tag{8.2}$$

$$H_2O \rightarrow \frac{1}{2}O_2 + 2H^+ + 2e^- \tag{8.3}$$

$$M + H_2O \rightarrow MO + 2H^+ + 2e^- \tag{8.4}$$

At the cathode, deposition of the metal M occurs through reduction of the metal:

$$M^{n+} + ne^- \rightarrow M \tag{8.5}$$

$$O_2 + H_2O + 4e^- \rightarrow 4OH^- \tag{8.6}$$

$$2H_2O + 2e^- \rightarrow H_2 + 2OH^- \tag{8.7}$$

At the anode, from Equations 8.2 through 8.4, the solution becomes more acidic as H^+ increases, while Equations 8.6 and 8.7 indicate the increase of hydroxyl (OH^-) ions, hence rendering the cathodic reaction more alkaline with respect to time.

In Section 8.2.5, a few case studies on potentiodynamic studies of lead-free solders are presented in order to gain some insight into the corrosion resistance of the various lead-free solder alloy formulations.

8.2.5 Selected Case Studies on Corrosion Resistance of Lead-Free Solders

8.2.5.1 Case Study 1: Sn-3.5Ag and Sn-0.8Cu

From Section 8.2, we are able to see that from the rest of the potential and current–potential curves, one will be able to analyse the tendency for dissolution in certain solder alloy formulations. For base metals which have a low rest potential, a lower rest potential will result in a higher tendency for dissolution. On the other hand, for noble metals which have a high rest potential, there is less tendency to form metal ions if its rest potential is higher. Furthermore, during the sweep from low potential to high potential, an abrupt rise in current density promotes the dissolution reaction. For example, Tanaka (2002) showed that in the case of the Sn-Ag and Sn-Cu alloys, the same potential was indicated for Sn-3.5Ag and Sn-0.8Cu, suggesting that the dissolution behaviour is dominated by the rest potential of the Sn. On the other hand, when comparing Sn-3.5Ag and Ag, the Sn element within the solder dominates the dissolution, as Sn-3.5Ag is unaffected by Ag. A similar observation was observed for Sn-0.8Cu. This evidence also implies that the intermetallics which form due to the addition of Ag and Cu into Sn (Ag_3Sn and Cu_6Sn_5, respectively) form stable compounds and do not dissolve in solution.

8.2.5.2 Case Study 2: Sn-3.0Ag-0.5Cu (SAC 305)/
Sn-1.0Ag-0.5Cu (SAC 105) Solder

Microgalvanic corrosion in SAC 305 solders was examined by Wang et al. (2012). They observed that faster cooling results in a microstructure which contains finer dendrites and microspheres with fibre-like Ag_3Sn IMCs. On the other hand, a faster cooling rate results in large, platelet-like Ag_3Sn. The different components of the solder solidify at different temperatures: β-Sn is the most difficult to solidify, while Ag_3Sn is the first to solidify at 209°C. Hence, below this temperature, Ag_3Sn IMCs begin to nucleate, leaving behind a solution increasingly rich in Sn and Cu. Next, below 205°C, Ag_3Sn and Cu_6Sn_5 nucleate. Hence, a fast cooling rate allows the precipitation of Ag_3Sn and Cu_6Sn_5 IMCs before the solidification of β-Sn, resulting in an overall finer morphology for the solder. On the other hand, a slower cooling rate allows the Ag_3Sn IMCs the possibility to coalesce into larger platelets.

An electrochemical cell is formed between Ag_3Sn, Sn and the moisture channel in the presence of moisture. Hence, microgalvanic corrosion occurs between the Ag_3Sn IMC (which acts as the cathode) and Sn (which is corroded). The platelets of larger surface areas are more susceptible to this microgalvanic corrosion.

8.2.5.3 Case Study 3: Doped SAC Solder Alloys

Hua and Yang (2011) studied doping of SAC solder alloys with In-Zn. They discovered through potentiodynamic polarisation studies that for the doping of Zn in excess of 1%, the corrosion was expedited compared with the pure SAC 305 solder. Additionally, the longest whisker was formed for the formulation 96.8(Sn-3.0Ag-0.5Cu)-0.2In-3Zn solder. From these findings, despite the mechanical advantages of the addition of In-Zn to SAC solder alloys, these works suggest that that these formulations may be as detrimental as microelectronics packaging. On a positive note, however, overall, the ECM in In-Zn-doped SAC 305 systems was less than that of the pure SAC 305 solder.

8.2.5.4 Case Study 4: Impact of Salt Exposure on Package Reliability

Liu et al. (2011) studied the effect of a 5% NaCl spray on SAC 305 wafer-level packages in terms of its long-term reliability and thermal fatigue. They found that the thermal cycling resistance for SAC 305 solders after being exposed to NaCl solutions was 43% less than that of Sn-Pb solders. More distinctly, Sn-Pb solders' thermal cyclic behaviour seemed impervious to exposure to salt solutions. These findings are in contradiction to the potentiodynamic studies where SAC 305 and Sn-Pb solders were compared, where SAC 305 showed lower passivation current densities and larger passivation domains than Sn-Pb solders. It is thought that these contradicting observations are due to the different failure mechanisms demonstrated in the thermal cycles, that is, crack location and propagation.

8.2.5.5 Case Study 5: Role of Passivation Layer on Material Surface

Nordin et al. (2015) studied the corrosion of (0.2–1.0 wt%) Al-added SAC 105 through potentiodynamic polarisation analysis. Referring to Figure 8.6 of the potentiodynamic curve for this formulation, while the pure SAC 105 continously reacts with the aggressive solution, SAC 105 with the addition of Al has already begun to passivate at an earlier-onset voltage. The Al-added SAC 105 eventually shows a notable passivation layer on the surface, which halts further corrosion. Hence, the doping of Al in the SAC solder serves as a corrosion mitigation strategy to the basic SAC 105 formulation.

8.2.6 Experimental Methodologies to Investigate Corrosion

In Section 8.2.5, the potentiodynamic studies carried out to investigate the galvanic corrosion of a range of lead-free solder alloys have been discussed at length. In this section, further details of other experimental methods which complement the potentiodynamic studies are presented. For example, surface oxide characterisation is usually carried out by Auger electron spectroscopy (AES), low-energy electron loss spectroscopy (LEELS) and Mossbauer techniques to elucidate the chemical information of oxides. Ellipsometry, X-ray emission, the gravimetric method and AES are used to measure oxide thickness. The impact of ECM is usually evaluated using environmental tests such as the thermal humidity bias (THB) test and highly accelerated stress test (HAST), or using normal conditions such as the water drop (WD) test, thin-electrolyte-layer (TEL) test and linear voltammetry.

8.2.6.1 Water Drop Test

He et al. (2011) observed the SAC 305 solder alloy under a THB test in order to evaluate its migration and deposition process. Medgyes et al. (2015) studied the ECM of lead-free microalloyed low-Ag-content solders using the WD tests in NaCl solution.

The WD test is carried out using an interdigitated comb electrode structure according to the IPC-B-24 test board, as shown in Figure 8.7. The solder paste was deposited according to this pattern, and a test droplet of NaCl was dropped onto the electrode to simulate the condition of seawater where $10\,V_{dc}$ was applied. The time to failure was recorded, and the dendrite formation during this experiment was observed *in situ*. The susceptibility of the lead-free alloys tested was evaluated as follows: SAC 305 > SAC 0807 > SAC 405 > SAC 0307. Of particular interest is that SAC 0807 and SAC 0307 have different ECM susceptibilities despite having similar compositions. This is

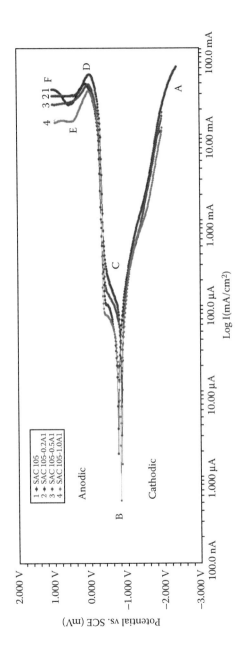

FIGURE 8.6
Potentiodynamic polarisation curve for SAC 105 and SAC 105-Al. (Reproduced from Nordin et al., 2015. With copyright permission. Published by Royal Society of Chemistry.)

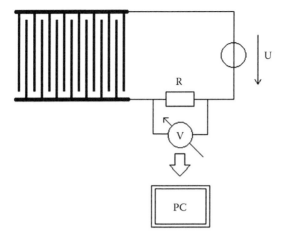

FIGURE 8.7
Schematic of the real-time measuring platform for WD tests. (Reproduced from Medgyes, B., et al., *Corrosion Science* 92: 43–47, 2015. With copyright permission. Published by Elsevier.)

due to the higher dissolution rate of the anode for SAC 0807, which correspondingly affects the higher corrosion rate.

8.2.6.2 ECM Investigation Using Voltage-Biased Microchannel

Minzari et al. (2011) have described in detail the experimental setup used to investigate the ECM and formation of the dendritic structure in tin solder alloys. They have utilised a microchannel structure to simulate the electrochemical 'cell' system between two metallic contacts and applied a bias of 5–12 V across these contacts. A time-lapse video was used to record *in situ* formation of the dendrites upon application of the voltage bias. Electron diffraction patterns have shown that the dendrite structure corresponds to that of metallic tin, irrespective of the magnitude of the voltage bias, but varies as a function of composition and surface morphology. This is shown in Figure 8.8. Additionally, an oxide layer was discovered to coat the metallic tin core along the growth direction of the dendrites. The ECM of tin across the channel is complex, as it involves the formation of strong pH gradients that arise in microvolumes within the channel due to the reactions across the microchannel electrodes. In particular, the hydrolysis of tin ions and water dissociation (creation of hydrogen ions) will cause the anode to be acidic, while the dissociation of water into hydroxyl ions at the cathode will cause the terminal to be alkaline. This scenario is further made complicated within the microvolumes due to local convection of the electrolyte and the kinetics of the electrochemical reactions. This *in situ* observation of the dendritic formation allows insight into the kinetics of the microgalvanic corrosion mechanism.

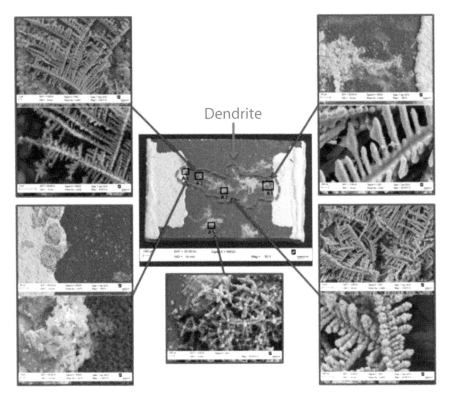

FIGURE 8.8
Scanning electron microscope images: 5 V sample at various sites of the dendrite (anode is on the left and cathode is on the right). (Reproduced from Minzari et al., 2011. With copyright permission. Published by Elsevier.)

8.3 Impact of Corrosion on Lead-Free Solder Performance

In the previous sections, the mechanism of corrosion in lead-free solders, particularly through galvanic corrosion, has been discussed. The effect of corrosion in board-level solder deposits is particularly significant with increased miniaturisation. Furthermore, it is generally thought that galvanic corrosion will be the major cause of failure in miniaturised circuits. For example, the rate of corrosion is measured in terms of micrometres of metal etched per year. This rate is constant for a specific combination of solder alloy and corrosive media, and it is regardless of the size of the solder alloy specimen. Thus, in miniaturised circuits, a small solder joint will be depleted at a faster rate than large solder joints. Furthermore, with smaller electrode gaps, a higher electric field is induced, and hence ECM is enhanced. This will eventually aggravate galvanic corrosion (Kawanobe and Otsuka, 1982; Nishigaki et al., 1986).

TABLE 8.1

Environmental Tests and Standards for Electronics Corrosion

Test	Standards	Uses
Atmospheric corrosion	IEC 60721-1	Environmental parameters and their severities
	IEC 60721-2	Environmental conditions appearing in nature
Humidity	IEC 68-2-3	Use and storage at high humidity
Thermal	IEC 68-2-1	Use and storage at low temperatures; adhesion of plating and coatings; physical distortion
	IEC 68-2-2	Use and storage at high temperatures; long-term ageing organics and oxidation of metals
Salt spray	IEC 68-2-11, ISO 9227, JESD 22-A107B	Compares resistance with deterioration from salt mist between specimens of similar materials or coatings
Soldering flux	J-STD-004A	Validates materials to produce the desired electrical and metallurgical interconnection; consequence of ECM and corrosion for flux usage

Additionally, there is less tolerance in terms of performance standards for miniaturised electronics packaging. For example, the JEDEC standard J-STD-004A requires that no filament growth should reduce the conductor spacing by more than 20%. This becomes more difficult to fulfil as the distance of contacts is reduced in miniaturised circuits. In terms of processing, the microsoldering process has a tendency to use no-clean fluxes in inert environments; that is, the use of these fluxes does not require a cleaning step postsoldering. However, the usage of these fluxes and consequent elimination of the cleaning step may increase ECM and corrosion. As a result, it will become detrimental to the overall circuit operation.

A more general summary of the industrial standards relevant to corrosion and the reliability factors relevant to the electronics packaging fabrication process is listed in Table 8.1. For example, the IEC standardised temperature and humidity categories which are recommended to be applied when determining the environmental classification of devices and components are IEC 60721-1 and IEC 60721-3-9.

Observation of failures due to corrosion include the manifestation of dendrites or pitting between conducting lines in a printed circuit board (PCB). An additional effect is when there is an absorption of moisture during the cathodic corrosion that results in a 'popcorn effect' which causes debonding and cracking.

8.4 Conclusion

In this chapter, the mechanism of corrosion of lead-free solders has been discussed. Due to the presence of dissimilar elements (i.e. tin and IMCs) in the lead-free solder microstructure, lead-free solders are susceptible to galvanic corrosion in the presence of humidity. This situation is further exacerbated in the presence of aggressive anions, introduced in marine environments or acid rain. The mechanism of galvanic corrosion was explained in terms of ECM, with the quantitative methods for determining the corrosion rate also elaborated upon. A few case studies of the corrosion behaviour of different lead-free solders have also been presented. Finally, the impact of corrosion on the lead-free solder performance in a microelectronics circuit operation was elaborated, with particular reference to industrial standards which assess the impact of corrosion on its performance, reliability and functionality.

References

Association Connecting Electronics Industries, IPC (2012). *Electronic Industry Market Data Update Report*. Bannockburn, Illinois. IPC. Retrieved from the IPC website: http://www.ipc.org/3.0_Industry/3.2_Market_Research/EIM-data-update/ Winter2012/ Market_Data_Update_2012_Winter.pdf

Abtew, M., and Selvaduray, G. (2000). Lead-free solders in microelectronics. *Materials Science and Engineering: R, Reports* 27(5–6): 95–141.

Ambat, R., Jellesen, M. S., Minzari, D., Rathinavelu, U., Johnsen, M. A., and Westermann, P. (2009). Solder flux residues and electrochemical migration failures of electronic devices. Presented at Proceedings of Eurocorr 2009, Nice, France.

He, X., Azarian, M., and Pecht, M. (2011). Evaluation of electrochemical migration on printed circuit boards with lead-free and tin-lead solder. *Journal of Electronic Materials* 40(9): 1921–1936.

Hua, L., and Yang, C. (2011). Corrosion behavior, whisker growth, and electrochemical migration of Sn–3.0Ag–0.5Cu solder doping with In and Zn in NaCl solution. *Microelectronics Reliability* 51(12): 2274–2283.

Johnsen, M. A. (2009). Effect of contamination on electrochemical migration. MSc thesis, Technical University of Denmark, Copenhagen.

Kawanobe, T., and Otsuka, K. (1982). Metal migration in electronic component. Paper presented at the *Proceeding of the 32nd Electronic Component Conference (USA)* San Diego, California.: 220–228. May 10–12, 1982.

Kotadia, H. R., Howes, P. D., and Mannan, S. H. (2014). A review: On the development of low melting temperature Pb-free solders. *Microelectronics Reliability* 54(6–7): 1253–1273.

Liu, B., Lee, T.-K., and Liu, K.-C. (2011). Impact of 5% NaCl salt spray pretreatment on the long-term reliability of wafer-level packages with Sn-Pb and Sn-Ag-Cu solder interconnects. *Journal of Electronic Materials* 40(10): 2111–2118.

Liu, X., Huang, M., Zhao, N., and Wang, L. (2014). Liquid-state and solid-state interfacial reactions between Sn–Ag–Cu–Fe composite solders and Cu substrate. *Journal of Materials Science: Materials in Electronics* 25(1): 328–337.

Medgyes, B., Horváth, B., Illés, B., Shinohara, T., Tahara, A., Harsányi, G., and Krammer, O. (2015). Microstructure and elemental composition of electrochemically formed dendrites on lead-free micro-alloyed low Ag solder alloys used in electronics. *Corrosion Science* 92: 43–47.

Minzari, D., Jellesen, M. S., Møller, P., Wahlberg, P., and Ambat, R. (2009). Electrochemical migration on electronic chip resistors in chloride environments. *IEEE Transactions on Device and Materials Reliability* 9: 392–402.

Minzari, D., Jellesen, M. S., Møller, P., & Ambat, R. (2011). On the electrochemical migration mechanism of tin in electronics. *Corrosion Science*, 53(10), 3366–3379.

Nishigaki, S., Fukuta, J., Yano, S., Kawabe H., Noda K., Fukaya M., Denjiyama M. (1986). A New Low Temperature Fireable Ag Multilayer Ceramic Substrate having Post-Fired Cu Conductor (LFC-2). Paper presented at the *Proceedings of the 4th International Society of Hybrid Microelectronics (Japan)Kobe*: 429–437. May 28–30, 1986.

Nordin, N. I. M., Said, S. M., Ramli, R., Sabri, M. F. M., Sharif, N. M., Arifin, N. A. F. N. M., & Ibrahim, N. N. S. (2014). Microstructure of Sn–1Ag–0.5Cu solder alloy bearing Fe under salt spray test. *Microelectronics Reliability*, 54(9–10), 2044–2047.

Tanaka, H. (2002). ESPEC Technology Report 14, 1.

Tompkins, H. G. (1973). The effect of alloy composition on the interaction of NO_2 with tin-lead alloys. *Surface Science* 39(1): 143–148.

Yu, D. Q., Jillek, W., & Schmitt, E. (2006). Electrochemical migration of lead free solder joints. *Journal of Materials Science: Materials in Electronics*, 17(3), 229–241.

9

From Deep Submicron Degradation Effects to Harsh Operating Environments: A Self-Healing Calibration Methodology for Performance and Reliability Enhancement

Eric J. Wyers and Paul D. Franzon

CONTENTS

ABSTRACT   In this chapter, we discuss recent calibration methodology developments for self-healing circuits and systems. Self-healing circuits and systems provide a means for performance and reliability enhancement in the presence of various degradation mechanisms, from deep submicron effects to harsh operating environments. The resulting self-healing calibration problem, however, may be difficult to solve with naive convex optimisation approaches, or may require relatively high complexity that comes along with global optimisation strategies. In this work, we adapt direct search optimisation algorithms to efficiently and effectively solve practical self-healing problems, and several test cases are used to demonstrate their efficacy. Future research directions are also discussed.

9.1 Introduction

The design of reliable, high-performance analogue, mixed-signal and radio frequency integrated circuits (RFICs) presents many challenges. Runtime performance and reliability can be degraded by variations in process, voltage, temperature and ageing (PVTA); impedance mismatches; and extreme operating conditions, such as radiation exposure and partial or total device- and block-level failures, all of which can lead to overdesigning the susceptible analogue blocks to cover the worst-case scenarios. An alternative to worst-case overdesign is to include on-chip digital calibration circuitry such that the susceptible blocks are made to be 'self-healing' [1]. Here, by self-healing, we mean those analogue, mixed-signal, and RFIC designs which cope with a plethora of performance and reliability degradations (extreme or otherwise) at runtime by the use of on-chip intelligence to 'heal' the circuit/ system back to an acceptable level of performance and reliable operation. In the literature, one finds an increasing amount of self-healing implementations being used to solve challenging design problems, such as self-healing phase-locked loop (PLL) strategies [2], a self-healing millimetre-wave power amplifier implementation [3], a self-healing RF amplifier methodology [4], a self-healing input-match technique for RF front-end circuitry [5] and self-healing RF downconversion mixers [6]. An important aspect of the self-healing methodology is that it is not limited to a particular circuit fabrication technology. For example, heterogeneous integration applications [7], where dissimilar fabrication technologies (e.g. GaN high-electron-mobility transistor [HEMT], InP heterojunction bipolar transistor [HBT] and complementary metal oxide semiconductor [CMOS]) are cointegrated in a common package to leverage the performance benefits of the respective individual technologies, can also greatly benefit from self-healing techniques. In this particular case, self-healing can be used to correct for degradations due to the mechanical stress impact of the heterogeneous integration environment on device electrical performance [8], or to mitigate performance and reliability degradation due to difficult-to-manage thermal hotspots [9].

The digitally assisted approach to complex analogue/RF calibration seeks to reduce the design complexity in the analogue domain through the use of compensation techniques in the digital domain. To motivate the use of digital correction techniques for analogue, mixed-signal and RF calibration problems, we consider, as an example, the particular case of RFIC block- and system-level components. Purely analogue-based calibration techniques, while popular for low-frequency applications, are not generally applicable in the RFIC case due to undesirable loading of the sensitive RF nodes by the calibration circuitry. Points in favour of digital integrated circuits, on the other hand, are that they support high integration densities and are more robust in the presence of PVTA variations and process technology scaling than their analogue/RF counterparts. In fact, digital calibration techniques for solving RFIC calibration problems have been employed in many challenging application spaces, such as wideband voltage-controlled oscillator (VCO) tuning range calibration [10] and input-referred second-order intercept point enhancement for an RF mixer [11]. Thus, incorporating digital calibration control (as opposed to purely analogue control), along with measurement sensors, an analogue-to-digital converter (ADC) and actuators, yields a self-healing calibration scheme which is robust and general in purpose and, with properly designed sensors, can mitigate undesirable loading effects for the blocks under calibration/healing [12].

In Figure 9.1, a block diagram of a digital self-healing calibration scheme is shown. Sensors are used to sense the current performance state (or 'health') of the circuit/system, and the actuators provide a means for performance and reliability correction to be made. It should be noted that to accommodate the calibration of multiple blocks or system-level components on an integrated circuit in an efficient manner, the digital calibration algorithm block and the ADC may be reused for the various calibration tasks. The digital calibration algorithm shown in Figure 9.1 takes as input the digitised sensor measurements and, by means of a properly defined objective function $f(x)$ relating to block- or system-level performance, adjusts the N-dimensional digital tuning knob vector x to make corrections to the circuit under healing until the desired performance is achieved. In a typical application, the goal of the self-healing algorithm is to determine an optimal N-dimensional digital tuning knob vector x_*, which sufficiently minimises the calibration objective f for a particular calibration problem. The self-healing calibration scheme depicted in Figure 9.1 can be viewed as a bound constrained and discretised optimisation problem. In this case, the digital tuning knobs are the optimisation variables. We say that these digital tuning knobs are bounded in that there is a lower and upper bound on each of the setting values. Further, the digital tuning knobs are only capable of making discrete corrections to the block or

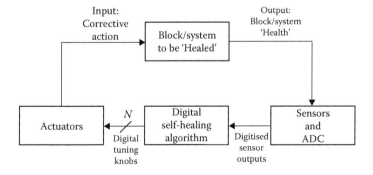

FIGURE 9.1
Conceptual block diagram of a digitally reconfigurable self-healing calibration scheme. (Adapted from Bowers, S. M., et al., *IEEE Trans. Microw. Theory Tech.* 61(3): 1301–1315, 2013.)

system being healed. Thus, we say that the self-healing calibration problem is bounded and discretised.

While the focus of this chapter is primarily on digitally reconfigurable self-healing algorithm techniques, sensing and actuation techniques also play a crucial role in enabling the self-healing approach to performance and reliability enhancement. Various sensor types can be employed in a self-healing application, such as phase sensors [13] and power detectors. On the actuator side, some examples include, among many others, digital-to-analogue converters (DACs) to, for example, control transistor biasing levels and switch control input signals which can be used to enable changes in the fundamental mode of operation [14] or place a specified number of integrated circuit components (e.g. transistors or capacitors) in parallel (or series) to vary circuit performance in some useful way [15]. Here, our main view on sensors and actuators is that so long as there are sensors which sense the current state (or health) of a circuit's performance, and actuators which provide a means for correction, any performance or reliability parameter of interest can be calibrated via the digitally reconfigurable self-healing calibration scheme depicted in Figure 9.1.

9.2 Motivation

Our approach to the digital self-healing algorithm seeks to provide an alternative to *ad hoc* techniques or techniques which effectively resort to an exhaustive search of the calibration space. *Ad hoc* self-healing techniques, such as that used in [16], can achieve self-healing functionality with relatively low digital overhead, but by definition, these techniques cannot be reused for multiple calibration tasks on an integrated circuit chip. Exhaustively searching the calibration space, such as that done in [17], is attractive for self-healing problems with a relatively low total number of digital calibration settings. For self-healing problems in relatively high dimensions, however, exhaustively searching the calibration space is prohibitive.

It is our view that the self-healing calibration problem for any block or system can be characterised by its problem dimension (i.e. N) and the calibration objective f. Based on this fundamental observation, our motivation is to develop a calibration algorithm strategy which satisfies the following: (1) It requires the circuit designer to specify only the calibration objective f, which can be sensed with appropriate sensors and corrected with appropriate actuators. (2) Once built, the calibration algorithm engine can be reused for the various calibration tasks. (3) The calibration algorithm should be capable of finding sufficiently optimal solutions in an efficient manner. Here, by 'efficient', we mean capable of finding optimal solutions by having to search only a small fraction of the total calibration space.

With the aforementioned desired calibration algorithm properties in mind, there are several classes of optimisation methods from which to choose. Probabilistic, or stochastic, methods, such as genetic algorithms [18], have been shown to be effective solvers of global optimisation problems [19]; however, probabilistic optimisation methods typically require a relatively large number of evaluations of the objective f to converge to an acceptable solution. Stochastic methods also typically possess a relatively large number of algorithm-specific parameters which must be tuned experimentally, for each particular problem, to achieve optimal performance [20]. Furthermore, it should be mentioned that the self-healing calibration problem is not a global optimisation problem: if a locally optimal solution in the response space satisfies the self-healing objective, then the locally optimal solution is sufficient. At the other 'extreme' on the optimisation spectrum, convex optimisation methods, such as those gradient descent–based methods mentioned in [21] and [22], leverage response convexity to locate optimal solutions in the response space. It is important to note, however, that response convexity for the self-healing calibration problem cannot be generally assumed.

9.3 Direct Search Algorithms for the Self-Healing Calibration Problem

Motivated by the observation that digitally reconfigurable analogue, mixed-signal, and RFIC designs often have locally optimal solutions which satisfy some arbitrary calibration objective, the overall goal of the research presented here is to provide the circuit designer with generally applicable calibration algorithm alternatives to the two extremes of convex and global optimisation approaches for noisy, and possibly nonconvex, self-healing calibration problems. To this end, we seek to apply a class of local optimisation algorithms, commonly referred to as direct search algorithms, to the self-healing calibration problem.

Direct search algorithms do not compute or approximate any derivatives of the objective function and are well suited for nonconvex, nonsmooth [23] and noisy [24] optimisation landscapes. These algorithms are also referred to as sampling methods in that the progression of the optimisation is controlled solely by the qualitative comparison of previously sampled points in the optimisation space. Thus, this minimal-in-complexity qualitative approach to optimisation leads to algorithms which are relatively straightforward to implement. Another advantage of direct search algorithms is that the relatively few algorithm-specific parameters can typically be set to default values such that optimisation simulations and experiments can commence quickly.

Direct search algorithms can be grouped into three categories [25]: (1) simplex methods, (2) pattern search methods and (3) methods with adaptive sets of search directions. Direct search algorithms have been in use since the late 1950s, and classical direct search algorithms from each of the three aforementioned categories include the Nelder–Mead simplex method [26], the Hooke–Jeeves pattern search method [27] and the Rosenbrock method with adaptive sets of search directions [28]. In what remains, we shall restrict our scope to a subset of those classical direct search algorithms, developed in the 1960s, which have been well documented, have been successfully applied to engineering optimisation problems and are relatively straightforward to implement. A key advantage of the classical variety of direct search algorithms is that they were successfully implemented on the relatively primitive computing technologies of the 1960s, suggesting that their time and space complexity requirements for self-healing calibration problems will be relatively low.

While the Nelder–Mead and Hooke–Jeeves algorithms were developed to be generally applicable to a broad range of problems, it is generally understood that the Rosenbrock method was developed primarily to solve problems with characteristics similar to those of a peculiar test problem, that is, the so-called Rosenbrock banana function [29], developed by Rosenbrock himself. Additionally, Powell observed that the Hooke–Jeeves algorithm is generally more successful than the Rosenbrock algorithm [30]. For these reasons, we choose to focus our attention exclusively on the application of the Nelder–Mead and Hooke–Jeeves algorithms to the self-healing calibration problem. Although our algorithmic focus will necessarily be constrained, interested readers may wish to broaden their scope and gain a better understanding of classical direct search algorithms from a historical standpoint. In particular, additional algorithms which may be of interest are the simplex methods of Spendley, Hext, and Himsworth [31] and M. J. Box [32]; the 'evolutionary operation' method of G. E. P. Box [33]; the pattern search algorithm commonly referred to as coordinate search [34,35]; and the multidirectional search (MDS) algorithm [36].

A majority of the remainder of this section is used to describe several key bounded and discretised algorithmic components for self-healing, and a self-healing strategy for low-dimensional calibration problems (i.e. problems with relatively low values of N). This section concludes with a proposed self-healing algorithm suited for high-dimensional calibration problems. To aid designers in incorporating our algorithms into their self-healing design flow, we mention here that MATLAB® implementations of the developed self-healing algorithms are freely available and may be found in [37].

9.3.1 Bounded and Discretised Nelder–Mead Algorithm

Here, we first give a brief description of the Nelder–Mead simplex algorithm, in its unmodified, unconstrained form. In contrast to the bounded and discretised self-healing case, by unconstrained, we mean those optimisation problems which pose no constraints on the optimisation variables. The goal in the unconstrained case is to use Nelder–Mead to minimise a real-valued function $f(x)$, $f:R^N \rightarrow R$, of N real decision variables belonging to vector x (we use R to denote the set of real numbers). For unconstrained problems, the components of vector x can take on any real number value, that is, $x \in R^N$. The Nelder–Mead algorithm maintains a simplex $S = \{x_1,..., x_{N+1}\}$ containing $N + 1$ vertices, where each vertex is an approximation to an optimal point in the unconstrained optimisation space R^N [24]. Conceptually speaking, the Nelder–Mead simplex S can be viewed as an $N \times N + 1$ matrix whose columns satisfy $x_i \in R^N$ for $i = 1,...,N$. The vertices (or columns) of simplex (or matrix) S are sorted according to their respective objective function values as

$$f(x_1) \le f(x_2) \le ... \le f(x_{N+1}) \tag{9.1}$$

The 'best point' in the simplex, having the lowest associated objective function value, is x_1, and the 'worst point' in the simplex, having the highest associated objective function value, is x_{N+1}. At each iteration, the unconstrained Nelder–Mead algorithm attempts to replace the worst point x_{N+1} by generating trial vertices $x(\mu) \in R^N$ of the form

$$x(\mu) = (1+\mu)\bar{x} - \mu x_{N+1} \tag{9.2}$$

where $\mu \in R$ is a coefficient associated with a particular step in the algorithm, and the centroid $\bar{x} \in R^N$ is given by $\bar{x} = (x_1 + ... + x_N)/N$. The steps in the Nelder–Mead algorithm which use Equation 9.2 to generate the trial vertices are referred to as the reflection, expansion, outside contraction and inside contraction steps, with respective coefficients μ_R, μ_E, μ_O and μ_I. In the event that the reflection, expansion, outside contraction or inside contraction steps fail to find an improvement on the worst point x_{N+1} in the simplex S, the algorithm performs a shrink step in which all of the vertices in S, except for the best point x_1, are replaced with shrink vertices $w_i \in R^N$ of the form

$$w_i = x_1 + \sigma(x_{i+1} - x_i) \tag{9.3}$$

for $i = 1,...,N$, and $\sigma \in R$ is the shrink coefficient. Typical values for the coefficients in Equations 9.2 and 9.3 are $\{\mu_R, \mu_E, \mu_O, \mu_I, \sigma\} = \{1, 2, 1/2, -1/2, 1/2\}$. Even though the best point x_1 may not be replaced with a better point in an iteration of the unconstrained Nelder–Mead algorithm, if the reflection, expansion, outside contraction or inside contraction steps are successful in replacing the worst point x_{N+1}, then the average simplex objective function value $\bar{f} \in R$, defined as [24]

$$\overline{f} = \frac{1}{N+1} \sum_{i=1}^{N+1} f(x_i),$$ (9.4)

will be reduced; however, in the event that a shrink step is taken by the algorithm, $\overline{f}$ in Equation 9.4 can increase.

The bounded and discretised Nelder–Mead (BDNM) algorithm, which we developed specifically for the self-healing calibration problem, requires several key modifications relative to the unconstrained Nelder–Mead algorithm for effective operation; a full listing of the BDNM algorithm can be found in [38]. The main modifications relate to the fact that the self-healing calibration problem is bounded and discretised. The self-healing problem is constrained to the set of bounded and discretised optimisation variables Ω, defined as $\Omega = \{x \in Z^N : L \le x \le U\}$, where Z denotes the integers, and $L, U \in Z^N$ are the N-dimensional vectors of lower and upper bounds on the digital tuning knobs, respectively. Since discretisation does not preserve linear independence of the simplex search directions, in our BDNM implementation, we make use of two simplices: one simplex, S, in which the vertices are constrained to the bounded and discretised set Ω, and the other simplex, $\hat{S}$, in which the vertices are bound constrained, but not discretised [38]. For clarity, we use the vertices of $\hat{S}$ to compute 'continuous' trial and shrink points, which are then discretised for inclusion into the bounded and discretised simplex S. We also use the vertices of $\hat{S}$ to test for convergence in the BDNM algorithm; in this way, the user does not need to specify a termination parameter, as is typically done when using the unconstrained Nelder–Mead algorithm [39]. In Figure 9.2, we show an example in two dimensions of a bounded and discretised simplex S with vertices x_1, x_2 and x_3, along with the various trial and shrink vertices.

Another key modification relates to the effective handling of a phenomenon that we discovered in the process of developing the BDNM algorithm: the bounded and discretised simplex vertices can repeat after a shrink step. That is, an iteration s of the BDNM algorithm which results in a shrink and produces a simplex S_s can lead to a later iteration t in which the produced simplex after a shrink step S_t has vertices such that $S_t = S_s$. It is our understanding that this phenomenon is not present in the unconstrained Nelder–Mead case, and the main problem arising from this scenario is that the BDNM algorithm is not generating new, possibly better, trial points, and when left uncorrected, BDNM is repeating simplex vertices which have already been evaluated, leading to an inefficient use of calibration resources. We correct this problem by incorporating a simplex cache of size D. The key idea here is that the cache stores simplices from the previous D iterations resulting in a shrink, and in the event that a shrink iteration occurs, the BDNM algorithm compares the resulting simplex with the cache contents, and if simplex repetition is detected, the BDNM algorithm is terminated. To further highlight this problem and our proposed solution, we present here an example

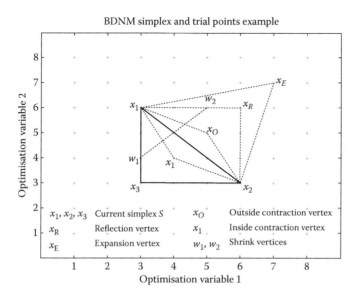

FIGURE 9.2
Two-dimensional example for the BDNM simplex and trial points.

in two dimensions. Figure 9.3 shows a randomly generated response surface; this is our example calibration objective response f, and Figure 9.4 shows the problem of simplex vertices repeating for the case of BDNM applied to the problem shown in Figure 9.3, with a simplex repetition cache length of $D = 0$; that is; simplex repetition detection is disabled. For clarity, in Figure 9.4 we show the iteration histories of the average simplex objective function value $\bar{f}$. As $\bar{f}$ can increase for those iterations resulting in a shrink, the repetition in simplex vertices due to shrinking can easily be seen in Figure 9.4, especially after iteration 10 of the BDNM algorithm. In Figure 9.5, we show the BDNM iteration histories for $\bar{f}$ with a simplex repetition cache length of $D = 1$. As shown in Figure 9.5, once the BDNM algorithm detects a repetition in simplex vertices resulting from a shrink step, the algorithm is terminated, eliminating an inefficient use of calibration resources.

9.3.2 Bounded and Discretised Hooke–Jeeves Algorithm

Prior to describing our bounded and discretised Hooke–Jeeves (BDHJ) algorithm developed for the self-healing problem, we first give a brief description of the Hooke–Jeeves algorithm in its unmodified form. The Hooke–Jeeves algorithm generates trial points $x_P \in R^N$ belonging to a 'pattern', and the trial points are of the form

$$x_P = x_C \pm h z_i, \tag{9.5}$$

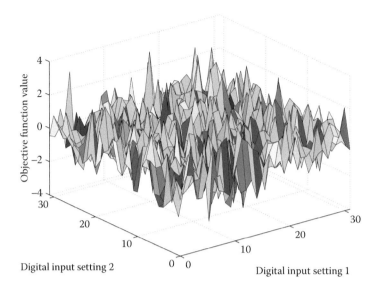

FIGURE 9.3
Example response surface for the simplex repetition phenomenon.

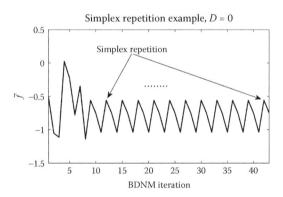

FIGURE 9.4
Example of simplex repetition due to shrinking in the BDNM algorithm.

where $x_C \in R^N$ is the current point, $h \in R$ is the pattern size and $z_i \in R^N$, $i = 1,...,N$, are the linearly independent search directions. The algorithm also makes use of the so-called pattern move step [24] in an attempt to find better solutions in the optimisation space. The algorithm continues to search for improvement, by either the generation of trial points with Equation 9.5 or pattern moves, until no further improvement is made, at which point the pattern size h is reduced and the algorithm begins the process all over again. The algorithm terminates once the pattern size is reduced to its minimum value and no further improvement in objective function value is found.

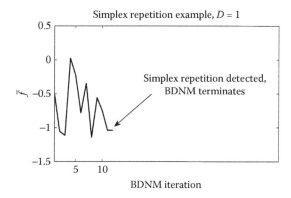

FIGURE 9.5
Example of simplex repetition detection in the BDNM algorithm.

The Hooke–Jeeves algorithm is well known in the optimisation community for solving both unconstrained problems and problems with simple boundary constraints [40]. To support boundary constraints, one typically implements the Hooke–Jeeves algorithm to discard trial points which leave the feasible optimisation region. Furthermore, the Hooke–Jeeves algorithm generates trial points by Equation 9.5, the points of which can conceptually be viewed as belonging to a discrete-point grid. Thus, unlike the Nelder–Mead algorithm, the Hooke–Jeeves algorithm extends with relative ease to the bounded and discretised self-healing optimisation problem. The main challenge to address is the manner by which the pattern size h is managed in the self-healing case. To solve this particular problem, we developed an approach which automatically generates the pattern sizes used within the BDHJ algorithm; no user involvement is required to fine-tune this process. Further details on our BDHJ implementation may be found in [41].

9.3.3 Bounded and Discretised Gradient Descent Algorithm

We now briefly describe the bounded and discretised gradient descent (BDGD) algorithm, developed in this work for comparative purposes. Gradient descent–based algorithms are widely used in the calibration of analogue integrated circuits, with three of the many recent examples found in [42–44]. As previously mentioned, convexity in the self-healing response cannot be generally assumed, and gradient descent–based techniques can perform very poorly when applied to nonconvex problems [38]. Nevertheless, due to their popularity in the analogue integrated circuit design community, it is still useful for us to mention the developed gradient descent–based calibration approach to put our direct search calibration algorithms into a proper context relative to the state of the art.

Our implementation of a gradient descent technique customised for the self-healing calibration problem most closely resembles the sign–sign least mean squares (SS-LMS) variant [45] of the LMS algorithm. In the SS-LMS algorithm, the polarities of the input regression and error terms are multiplied to produce a gradient estimate. However, the self-healing calibration scheme under consideration here, depicted in Figure 9.1, has no explicit means of generating the regression and error terms required by the SS-LMS algorithm. Nevertheless, a gradient estimate about any point in the self-healing response space can be produced by searching the unit coordinate directions about the point under consideration. In fact, based on this observation, the BDGD algorithm can be viewed as a pattern search algorithm with a fixed pattern size h. Thus, our implementation of the BDGD algorithm essentially resembles that of the BDHJ algorithm, with the exception that the BDGD algorithm does not perform any pattern moves and the pattern size does not change. A full description of our BDGD algorithm implementation may be found in [38].

While it has been our experience that the BDGD algorithm is not a very effective solver for challenging nonconvex self-healing problems, we should point out that for convex (or nearly convex) problems, the BDGD algorithm may be an option worth considering. For example, for a one-dimensional voltage-controlled delay line (VCDL) problem, where the objective was to heal the effects of negative-bias temperature instability (NBTI) on VCDL operation, BDGD was shown to work satisfactorily [46].

9.3.4 Limitations of Direct Search Algorithms

For the unconstrained optimisation problem, rather than attempting to obtain the global optimum of objective function f, direct search algorithms are instead often applied in an attempt to find a locally optimal solution of f. The globally optimal solution of f is a solution in R^N which yields the lowest value of the objective. On the other hand, a locally optimal solution of f (often referred to as a local minimiser, or, more simply, a minimiser) is a solution $x_* \in R^N$ which satisfies $f(x_*) \leq f(x)$ for all $x \in R^N$ near x_*. A commonly used definition for 'near', in the locally optimal sense, is all points x which satisfy $\|x - x_*\| < \varepsilon$, where $\| \cdot \|$ denotes the Euclidean, or l^2, norm, and ε is a positive scalar dictating the 'radius' over which solution x_* is considered locally optimal.

A first-order necessary condition for a point to be considered a local minimiser is that it must be a stationary point [24]. In the unconstrained case, a stationary point $x \in R^N$ is such that the gradient of f evaluated at x, $\nabla f(x) \in R^N$, is zero valued. Under general conditions, the unconstrained implementations of the direct search algorithms under investigation in this work are not guaranteed to converge to a minimiser. The Hooke–Jeeves algorithm is, however, provably convergent to a stationary point [47]. It should be noted, though, that a stationary point is not necessarily a minimiser.

Unlike the Hooke–Jeeves algorithm which is guaranteed to converge to a stationary point, the convergence theory available for the unconstrained

Nelder–Mead algorithm is specific to certain classes of optimisation problems. Convergence results for Nelder–Mead are given in [48], but are applicable to one- and two-dimensional problems with strictly convex objective functions. As mentioned in [49], Nelder–Mead can converge to nonminimisers, and in [50], the Nelder–Mead algorithm is shown to converge to a nonstationary point with a family of convex functions in two dimensions. Despite having limited or, in some cases, unfavourable convergence theory results, the Nelder–Mead algorithm has, nevertheless, been shown to work well for a wide range of problems encountered in practice [48].

In addition to the available convergence theory results for the algorithms under consideration in this work, we are also particularly interested in the limitations of direct search algorithms with respect to the effect of problem dimensionality on performance. The performance of direct search algorithms is known to deteriorate as the dimensionality of the problem (i.e. the value of N) increases [51], a phenomenon known as the 'curse of dimensionality'. As mentioned in [37], direct search algorithms for unconstrained problems are suited for problems of $N = 30$ or fewer optimisation variables, with 10 or fewer optimisation variables being ideal.

9.3.5 Bounded and Discretised Neighbourhood Search Algorithm

Self-healing calibration problems come in two varieties: target calibration problems and blind calibration problems. In the target calibration problem, we seek to find a solution x_* which satisfies $f(x_*) \leq f_t$, where $f_t \in R$ is the calibration target performance level. In the blind calibration case, that is, problems for which setting a numerical calibration target is difficult or infeasible, we seek to obtain a solution which satisfies some local optimality condition.

The developed BDNM and BDHJ direct search algorithms (and BDGD, which was developed for comparative purposes) are not guaranteed to find a locally optimal solution, in either the target or the blind sense. Thus, to guarantee that an optimal solution is obtained during self-healing, we developed the bounded and discretised neighbourhood search (BDNS) algorithm [38]; however, the BDNS approach to local optimisation that we advocate here is not suitable for high-dimensional problems. In basic terms, the BDNS-based strategy for low-dimensional problems is to, first, apply the chosen direct search algorithm to the problem and, second, search the local neighbourhood about the solution obtained by the chosen direct search algorithm, until a solution is found which either satisfies the target objective (for target problems) or is locally optimal (for blind problems). Figure 9.6 shows a flow chart of the proposed self-healing calibration strategy for low dimensions. The key idea here is that we use the chosen direct search algorithm to find an intermediate solution with high efficiency, and use the less efficient BDNS algorithm only to ensure that the final solution is locally optimal in some sense. When we use BDNM in conjunction with BDNS, as shown in Figure 9.6, we refer to this as BDNM-BDNS, and similarly for BDHJ and BDGD.

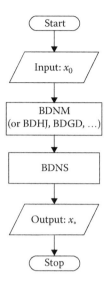

FIGURE 9.6
Proposed self-healing strategy for low-dimensional problems.

Figure 9.7 shows an example of applying the BDNS algorithm for a target calibration problem in two dimensions. In Figure 9.7, the starting point for BDNS is the solution obtained by the chosen direct search algorithm, x_0. At each iteration of BDNS, the algorithm searches neighbourhood subsets about the initial iterate x_0 for solutions which satisfy the target objective f_t. In the example of Figure 9.7, during the first iteration, eight points are evaluated, and the objective function evaluated at the best point found, $f_1 = f(x_1)$, does not satisfy the target goal. In the second iteration, 16 points are evaluated, and the best point found, x_2, also does not satisfy the target. Finally, in the third iteration, 24 additional evaluations of the objective f are made, and a solution, x_3, is found, having an objective function value $f_3 = f(x_3)$, which satisfies $f_3 < f_t$. For blind problems, BDNS proceeds in a similar manner [38], with the difference being that the user defines the locally optimal neighbourhood size.

From the example of Figure 9.7, counting the objective function evaluations taken at each iteration, we see that BDNS takes a total of 48 evaluations of the objective before finding a solution which satisfies the target. Thus, it is clear that such a local search by BDNS is impractical for self-healing calibration problems in high dimensions. For low-dimensional problems, however, the upside to applying BDNS is that the algorithm is guaranteed to find an optimal solution for target calibration problems (so long as such a solution exists in the calibration response space), and for blind problems, the local optimality criterion is user programmable to accommodate a wide range of self-healing calibration needs.

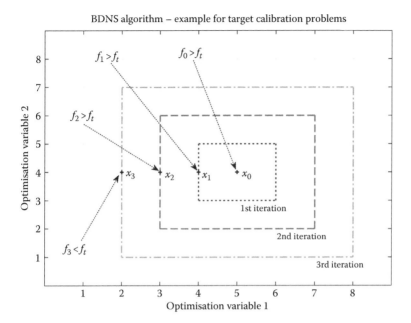

FIGURE 9.7
BDNS calibration example for target calibration problems in two dimensions.

9.3.6 Hybrid Direct Search Self-Healing Algorithm for High-Dimensional Problems

The use of the BDNS algorithm to ensure final solution optimality can be costly at higher dimensions. However, for high-dimensional problems, if one is not using the BDNS algorithm in conjunction with a bounded and discretised direct search algorithm, there is no guarantee that the chosen direct search algorithm will obtain an optimal solution upon termination. One way to address this problem is to simply restart the chosen calibration algorithm in the hopes of finding a better solution on the next attempt. In fact, in our initial experimentations, we developed so-called restarted versions of the BDNM, BDHJ and BDGD algorithms, referred to as BDNMR, BDHJR and BDGDR, respectively, where, for example, BDNMR stands for 'bounded and discretised Nelder–Mead with restart'. To be clear, when an algorithm is restarted, it uses the best point obtained in the previous iteration as the starting point x_0 for the next iteration.

Taking the above concept a bit further, we now introduce our hybrid restart algorithm which comprises the developed BDNM and BDHJ algorithms [41]. To motivate this development, we consider the trial point generation characteristics of the individual BDNM and BDHJ algorithms. The BDNM algorithm generates trial points based on the simplex vertices, and as the BDNM simplex changes at each iteration, BDNM generates trial points with search direction variety, and such variety is much needed at relatively high problem dimensions. However, the BDNM simplex can 'collapse' and become singular during the optimisation process [38], affecting the algorithm's ability to search in directions which span the optimisation space. On the other hand, BDHJ incorporates search directions which are guaranteed to remain linearly independent during the course of optimisation; however, the search directions used by the BDHJ algorithm remain fixed, lacking the variety offered by BDNM.

Our proposed hybrid direct search algorithm, the bounded and discretised Nelder–Mead and Hooke–Jeeves with restart (BDNMHJR) algorithm, was developed to leverage the search direction attributes of the individual BDNM and BDHJ algorithms. We should note that the Nelder–Mead and Hooke–Jeeves algorithms have been hybridised previously, such as in the 'Simpat' implementation [52]. The drawback to the Simpat approach, in particular, is that it requires the user to customise the algorithm's implementation; in our hybrid implementation, no user customisation is required. In Figure 9.8, we show a conceptual block diagram of the BDNMHJR algorithm. As shown, the

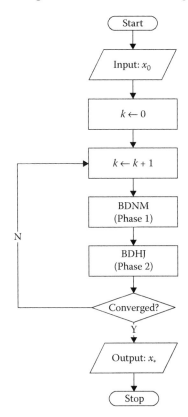

FIGURE 9.8
Conceptual block diagram of the BDNMHJR algorithm.

algorithm consists of applying the BDNM and BDHJ algorithms sequentially in two phases: the first phase consists of applying BDNM to the self-healing problem, and the second phase of applying BDHJ. Once an iteration k completes, the algorithm checks for convergence, which is defined differently for target and blind problems. For target problems, BDNMHJR takes as input the target goal f_t and terminates once a solution is found to meet the target, or if k exceeds the maximum allowed iterations $kmax$. Our recommended value for $kmax$ is 100 [41]. For blind problems, a parameter, $smax$, is passed to the algorithm, and in this case, $smax$ dictates the maximum allowed number of successive unsuccessful BDNMHJR iterations. An unsuccessful iteration is one in which the algorithm fails to improve on the best solution obtained thus far. The need for a parameter which sets the maximum allowed number of successive unsuccessful iterations appears in many fields of study, including a digital circuit optimisation application [53] and a host of other applications, such as in [54–58], and the value chosen for the equivalent $smax$ parameter in all of these applications is 10. Thus, when using BDNMHJR for blind self-healing problems, we recommend setting the $smax$ parameter to 10.

9.4 Experimental Results

We now present experimental results obtained for several complex self-healing calibration test cases. For the chosen test cases, we focus on several blocks used within an RF receiver front end. In Figure 9.9, we show such an RF receiver implementation that is suitable for a phased-array system [59]. The RF input is fed to a low-noise amplifier (LNA) block, and the goal is to downconvert the RF input signal to baseband. The PLL [60] is used to generate low-jitter signals to aid in the downconversion task and generate local oscillator (LO) signals, consisting of both in-phase (I) and quadrature (Q) types. The phase rotator (PR) block [61] provides phase selectivity in the signal conversion from an intermediate frequency (IF) to baseband.

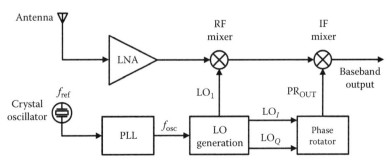

FIGURE 9.9
RF receiver front-end block diagram for a phased-array system.

We have obtained experimental self-healing results for the PLL and PR blocks, which will be presented in turn. Additionally, self-healing overhead estimates for those algorithms discussed next have been obtained, and reveal that the developed algorithms require relatively low overhead compared with the analogue/RF blocks being healed. We point interested readers to [37,38,41] for further details on the required overhead for the developed digital self-healing algorithms.

9.4.1 Two-Dimensional PLL Self-Healing Results

In Figure 9.10, we show a conceptual block diagram of the PLL used as a self-healing test case in this work [38]. The PLL test case under consideration is a two-dimensional problem, which is fabricated in 65 nm CMOS, and similar to the PLL implementation described in [62]. The two digital tuning knobs, $v_{trig} \in [1, 60]$ and $v_{ichg} \in [0, 255]$, control the timing and amplitude of the correction signal, and are used to minimise the voltage ripple on the VCO control line. The ultimate self-healing goal for the PLL is to reduce spurious tone levels in the output signal spectrum. The ideal output of the PLL is a 12 GHz sinusoid, and the spurious tones appearing at the PLL output can be caused by any number of degradation mechanisms, such as PVTA and radiation effects [63]. The PLL self-healing calibration problem presented here is blind, since the goal is to reduce the spurious tone content as much as possible. Figure 9.11 shows the measured PLL response [38], which is the detected peak voltage on the VCO control line, and our goal is to minimise the detected peak. The PLL response is characterised by many local minima, most of which are suboptimal.

We applied BDNM-BDNS, BDHJ-BDNS and BDGD-BDNS to the two-dimensional PLL calibration problem. BDHJ-BDNS and BDGD-BDNS did not make much improvement beyond the objective function value obtained at the initial iterate x_0, and thus they were not effective solvers of the PLL

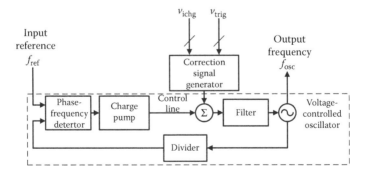

FIGURE 9.10
Conceptual self-healing PLL block diagram. (Adapted from Wyers, E. J., et al., *IEEE Trans. Circuits Syst. I Reg. Papers* 60(7): 1787–1799, 2013.)

Measured PLL response

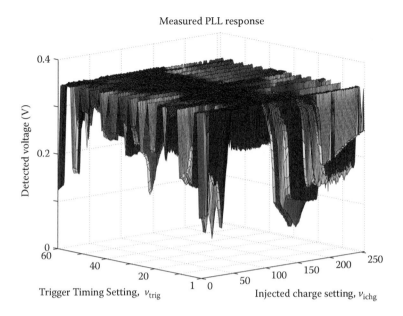

FIGURE 9.11
Measured PLL response. (Adapted from Wyers, E. J., et al., *IEEE Trans. Circuits Syst. I Reg. Papers* 60(7): 1787–1799, 2013.)

self-healing calibration problem. We point interested readers to [37,38] for a summary of the results obtained with BDHJ-BDNS and BDGD-BDNS, respectively. On the other hand, BDNM-BDNS was able to achieve significant reduction in the output spurious tone level, and in Figure 9.12, we show the reduction in the first harmonic reference spurs, at which BDNM-BDNS achieves greater than 10 dBc reduction in spurious power, which is typical of the performance obtained with several test chips. On average, BDNM-BDNS only required about 80 evaluations of the objective function to achieve an optimally healed PLL state [37] compared with the total 15,360 digital tuning knob setting possibilities, suggesting that BDNM-BDNS is not only a highly effective solver of the PLL self-healing problem, but also highly efficient. We should point out that BDNMHJR was applied to the PLL calibration problem in simulation as well, and like BDNM-BDNS, BDNMHJR was also able to effectively heal the PLL; however, as one might expect, BDNM-BDNS was more efficient than BDNMHJR for this relatively low-dimensional self-healing problem [37].

9.4.2 Eight-Dimensional PR Self-Healing Results

In Figure 9.13, we show a conceptual block diagram of the PR block used as a high-dimensional self-healing test case in this work [41]. The PR test case under consideration is an eight-dimensional problem, which is fabricated in

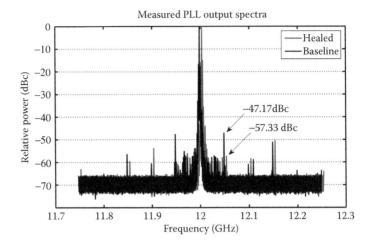

FIGURE 9.12
Self-healing results from applying BDNM-BDNS to the self-healing PLL problem. The spectrum after healing has been shifted by +5 MHz in frequency for clarity. (Adapted from Wyers, E. J., et al., *IEEE Trans. Circuits Syst. I Reg. Papers* 60(7): 1787–1799, 2013.)

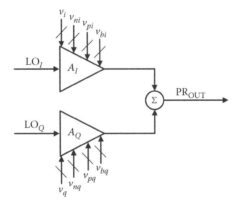

FIGURE 9.13
Conceptual self-healing PR block diagram. (Adapted from Wyers, E. J., et al., *IEEE Trans. Very Large Scale Integr. (VLSI) Syst.* 24(3): 1151–1164, 2016.)

45 nm silicon-on-insulator (SOI) CMOS and is equipped with self-healing circuitry to cope with aggressively scaled technology effects [64] and hot-carrier injection (HCI) ageing effects [65]. In total there are eight digital tuning knobs: four knobs, v_{pi}, v_{ni}, v_{pq}, $v_{nq} \in [0, 255]$, which are used to control the input bias level to variable-gain amplifiers (VGAs) A_I and A_Q; two knobs, v_{bi}, $v_{bq} \in [0, 255]$, which control the VGA currents; and two knobs, v_i, $v_q \in [0, 63]$, which are used to adjust the VGA gains. The PR self-healing objective is to

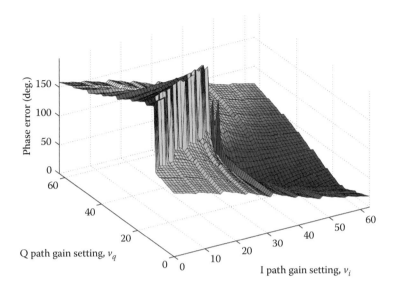

FIGURE 9.14
Simulated PR phase error response for a phase target of 236.25°.

find an optimal setting for the eight digital tuning knobs which produces a certain phase and gain at the PR output. The phase is user programmable over the full 360° range, and the gain is to be set at some constant level, say, 0 dB. The PR problem is a target problem, and the multiobjective goal is to achieve the target phase and gain within 2° and 0.5 dB error margins, respectively. Figure 9.14 shows the simulated PR phase error response for a target phase of 236.25°, and Figure 9.15 shows the simulated PR gain error response for a target gain of 0 dB.

We applied the BDNMR, BDHJR and BDGDR algorithms to the eight-dimensional PR problem in simulation, and all three of these algorithms were not able to achieve the gain and phase error requirements simultaneously [41]. The hybrid BDNMHJR algorithm was, however, shown to be an effective solver in both simulation and measurements obtained with several test chips. To show that both BDNM and BDHJ make effective contributions to the proposed hybrid BDNMHJR self-healing algorithm strategy for high-dimensional problems, we show in Figure 9.16 measured results for which BDNM was responsible for obtaining an optimal solution, and in Figure 9.17 we show measured results for which BDHJ was responsible for obtaining an optimal solution. On average, over all test chips and user-programmable phases, BDNMHJR required 520 evaluations of the objective function, which, when compared with the 2^{60} total number of digital setting possibilities in the self-healing response space, suggests highly efficient self-healing operation for this challenging eight-dimensional problem [41].

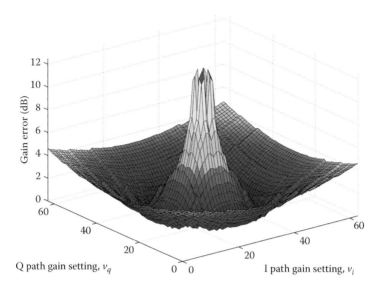

FIGURE 9.15
Simulated PR gain error response for a gain target of 0 dB.

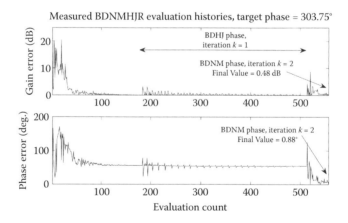

FIGURE 9.16
Measured BDNMHJR self-healing iteration histories showing optimal solution found during the BDNM phase.

9.5 Conclusions and Future Work

Self-healing is an effective tool for achieving high performance and enabling reliable operation for analogue, mixed-signal, and RFIC designs in the presence of myriad degradations and harsh operating conditions. In this work,

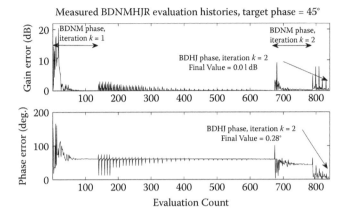

FIGURE 9.17
Measured BDNMHJR self-healing iteration histories showing optimal solution found during the BDHJ phase.

we have shown that direct search algorithms are an effective and efficient solver of challenging self-healing problems. Depending on the dimensionality of the self-healing problem, several strategies were introduced to give the circuit designer much-needed alternatives for those particular self-healing problems encountered in practice. Several complex test cases were undertaken for performance validation purposes, and the measured results confirmed that our proposed self-healing calibration strategies are robust and generally applicable.

On the algorithmic side of self-healing, several items remain to be investigated. First, to better understand the performance limitations of the developed algorithms, convergence properties and theoretical performance guarantees need to be established. Second, while we are generally quite pleased with the performance obtained by the algorithms chosen for this research, we are also interested in applying additional algorithms to self-healing problems to better understand relative performance differences and complexity requirements. Some algorithmic possibilities include relatively more modern direct search algorithms, such as the algorithm proposed in [66], genetic algorithms, particle swarm optimisation [67], the DIRECT algorithm [68] and possibly others of the stochastic variety, such as the simulated annealing algorithm [69]. Also, while the test cases in this work are shown to be challenging self-healing problems, we seek to apply our developed healing algorithms to additional self-healing block- and system-level implementations to uncover possible performance limitations which could motivate further modifications to the algorithms for more robust operation.

We also see the need to develop effective self-healing strategies which simultaneously address challenges on both the design side and the algorithmic side. One motivating example is to consider the matter of self-healing digital tuning knob selection. In this work, it was assumed that designers use some prior knowledge of the design to determine which knobs to add and what their lower and upper bounds should be. However, to achieve a truly robust self-healing implementation, a balance needs to be achieved: selecting 'too few' knobs may render the self-healing problem unsolvable, and on the other hand, adding 'too many' knobs may lead to a problem which is difficult to solve from a 'curse-of-dimensionality' perspective. From this particular example, one sees that the design side and the algorithmic side of self-healing are inextricably linked, and a more robust self-healing flow can be achieved if it is determined how to best simultaneously optimise these two important facets of self-healing.

Another item requiring further research is that related to self-healing algorithm performance validation prior to chip tape-out. In our work, we have made use of simulations wherever possible to validate the performance of our algorithms; however, it is not always feasible to simulate large portions of the response space to ensure performance, especially for high-dimensional self-healing problems. It is our view that the incorporation of machine learning techniques [70] to build representative self-healing implementation behavioural models, at both the block and system levels, will allow for relatively quick self-healing performance validation when compared with relying on expensive circuit simulations alone. Due, in part, to advances in the design optimisation state of the art, for example, recent work for the design optimisation of a highly efficient power amplifier [71], Bayesian optimisation [72], a supervised machine learning technique, may prove to be useful in generating relatively inexpensive models for self-healing algorithm performance validation. Surrogate modelling [73], also a supervised machine learning technique, has also been previously applied to behavioural modelling of analogue circuits [74], and could be a valuable tool for self-healing performance validation as well. Machine learning techniques may also be useful at self-healing design optimisation time for reducing the required complexity of the self-healing implementation and for digital tuning knob selection. Furthermore, several recent works, such as an indirect self-healing performance sensing scheme [75] and the use of 'side information' (e.g. data obtained prior to chip tape-out from simulations or from postfabrication measurements) for producing accurate analogue/RF block performance models [76], suggest that the incorporation of machine learning techniques will have a profound impact on future self-healing circuits and systems, and deserves further attention from the self-healing research community.

References

1. C. Maxey, G. Creech, S. Raman, et al. 2012. Mixed-Signal SoCs with In Situ Self-Healing Circuitry. *IEEE Des. Test Comput.* 29(6): 27–39.

2. W.-H. Chen and B. Jung. 2010. Self-Healing Phase-Locked Loops in Deep-Scaled CMOS Technologies. *IEEE Des. Test Comput.* 27(6): 18–25.

3. S. M. Bowers, K. Sengupta, K. Dasgupta, B. D. Parker, and A. Hajimiri. 2013. Integrated Self-Healing for mm-Wave Power Amplifiers. *IEEE Trans. Microw. Theory Tech.* 61(3): 1301–1315.

4. A. Goyal, M. Swaminathan, A. Chatterjee, D. C. Howard, and J. D. Cressler. 2012. A New Self-Healing Methodology for RF Amplifier Circuits Based on Oscillation Principles. *IEEE Trans. Very Large Scale Integr. (VLSI) Syst.* 20(10): 1835–1848.

5. T. Das, A. Gopalan, C. Washburn, and P. R. Mukund. 2005. Self-Calibration of Input-Match in RF Front-End Circuitry. *IEEE Trans. Circuits Syst. II Exp. Briefs* 52(12): 821–825.

6. A. Goyal, M. Swaminathan, and A. Chatterjee. 2009. Self-Calibrating Embedded RF Down-Conversion Mixers. In *Proceedings of IEEE Asian Test Symposium (ATS)*, pp. 249–254.

7. T. R. Harris, G. Pavlidis, E. J. Wyers, et al. 2016. Thermal Raman and IR Measurement of Heterogeneous Integration Stacks. In *Proceedings of IEEE Intersociety Conference on Thermal and Thermomechanical Phenomena in Electronic Systems (ITherm)*, Las Vegas, NV, May 31–June 3, pp. 1505–1510.

8. E. J. Wyers, T. R. Harris, W. S. Pitts, J. Massad, and P. D. Franzon. 2015. Characterization of the Mechanical Stress Impact on Device Electrical Performance in the CMOS and III-V HEMT/HBT Heterogeneous Integration Environment. In *Proceedings of IEEE International 3D Systems Integration Conference (3DIC)*, Sendai, Japan, August 31–September 2, pp. 1–4, TS8.27.1–TS8.27.4.

9. T. R. Harris, E. J. Wyers, L. Wang, et al. 2015. Thermal Simulation of Heterogeneous GaN/InP/Silicon 3DIC Stacks. In *Proceedings of IEEE International 3D Systems Integration Conference (3DIC)*, Sendai, Japan, August 31–September 2, pp. 1–4, TS10.2.1–TS10.2.4.

10. A. Aktas and M. Ismail. 2004. CMOS PLL Calibration Techniques. *IEEE Circuits Devices Mag.* 20(5): 6–11.

11. S. Rodriguez, A. Rusu, L.-R. Zheng, and M. Ismail. 2008. CMOS RF Mixer with Digitally Enhanced IIP2. *Electron. Lett.* 44(2): 121–122.

12. J. Wilson and M. Ismail. 2009. Input Match and Load Tank Digital Calibration of an Inductively Degenerated CMOS LNA. *Integration VLSI J.* 42(1): 3–9.

13. H. Wang, K. Dasgupta, and A. Hajimiri. 2011. A Broadband Self-Healing Phase Synthesis Scheme. In *Proceedings of IEEE Radio Frequency Integrated Circuits Symposium (RFIC)*, Baltimore, MD, June 5–7, pp. 1–4.

14. N. van Bavel, E. J. Wyers, and J. J. Paulos. 2003. Output Driver for High Speed Ethernet Transceiver. U.S. Patent 6,665,347.

15. E. J. Wyers, D. Stiurca, and J. J. Paulos. 2006. On-Chip Calibrated Source Termination for Voltage Mode Driver and Method of Calibration Thereof. U.S. Patent 7,119,611.

16. D. C. Howard, P. K. Saha, S. Shankar, et al. 2014. A SiGe 8–18-GHz Receiver with Built-in-Testing Capability for Self-Healing Applications. *IEEE Trans. Microw. Theory Tech.* 62(10): 2370–2380.

17. K. Jayaraman, Q. Khan, B. Chi, W. Beattie, Z. Wang, and P. Chiang. 2010. A Self-Healing 2.4 GHz LNA with On-Chip S11/S21 Measurement/Calibration for In-Situ PVT Compensation. In *Proceedings of IEEE Radio Frequency Integrated Circuits Symposium (RFIC)*, Anaheim, CA, May 23–25, pp. 311–314.

18. P. Fernando, S. Katkoori, D. Keymeulen, R. Zebulum, and A. Stoica. 2010. Customizable FPGA IP Core Implementation of a General-Purpose Genetic Algorithm Engine. *IEEE Trans. Evol. Comput.* 14(1): 133–149.

19. M. Srinivas and L. M. Patnaik. 1994. Genetic Algorithms: A Survey. *Computer* 27(6): 17–26.

20. D. W. Boeringer, D. H. Werner, and D. W. Machuga. 2005. A Simultaneous Parameter Adaptation Scheme for Genetic Algorithms with Application to Phased Array Synthesis. *IEEE Trans. Antennas Propag.* 53(1): 356–371.

21. B. Widrow and M. A. Lehr. 1990. 30 Years of Adaptive Neural Networks: Perceptron, Madaline, and Backpropagation. *Proc. IEEE.* 78(9): 1415–1442.

22. B. Widrow and M. Kamenetsky. 2003. Statistical Efficiency of Adaptive Algorithms. *Neural Netw.* 16(5–6): 735–744.

23. T. G. Kolda, R. M. Lewis, and V. Torczon. 2003. Optimization by Direct Search: New Perspectives on Some Classical and Modern Methods. *SIAM Rev.* 45(3): 385–482.

24. C. T. Kelley. 1999. *Iterative Methods for Optimization.* Frontiers in Applied Mathematics. Philadelphia: Society for Industrial and Applied Mathematics.

25. R. M. Lewis, V. Torczon, and M. W. Trosset. 2000. Direct Search Methods: Then and Now. *J. Comput. Appl. Math.* 124(1–2): 191–207.

26. J. A. Nelder and R. Mead. 1965. A Simplex Method for Function Minimization. *Comput. J.* 7(4): 308–313.

27. R. Hooke and T. A. Jeeves. 1961. "Direct Search" Solution of Numerical and Statistical Problems. *J. ACM* 8(2): 212–229.

28. H. H. Rosenbrock. 1960. An Automatic Method for Finding the Greatest or Least Value of a Function. *Comput. J.* 3: 175–184.

29. J. W. Bandler, A. S. Mohamed, M. H. Bakr, K. Madsen, and J. Sondergaard. 2002. EM-Based Optimization Exploiting Partial Space Mapping and Exact Sensitivities. *IEEE Trans. Microw. Theory Tech.* 50(12): 2741–2750.

30. M. J. D. Powell. 1970. A Survey of Numerical Methods for Unconstrained Optimization. *SIAM Rev.* 12(1): 79–97.

31. W. Spendley, G. R. Hext, and F. R. Himsworth. 1962. Sequential Application of Simplex Designs in Optimisation and Evolutionary Operation. *Technometrics* 4(4): 441–461.

32. M. J. Box. 1965. A New Method of Constrained Optimization and a Comparison with Other Methods. *Comput. J.* 8: 42–52.

33. G. E. P. Box. 1957. Evolutionary Operation: A Method for Increasing Industrial Productivity. *Appl. Statist.* 6(2): 81–101.

34. V. Torczon. 1997. On the Convergence of Pattern Search Algorithms. *SIAM J. Optim.* 7(1): 1–25.

35. N. J. Higham. 1993. Optimization by Direct Search in Matrix Computations. *SIAM J. Matrix Anal. Appl.* 14(2): 317–333.

36. V. Torczon. 1991. On the Convergence of the Multidimensional Search Algorithm. *SIAM J. Optim.* 1(1): 123–145.

37. E. J. Wyers. 2013. Direct Search Calibration Algorithms for Digitally Reconfigurable Radio Frequency Integrated Circuits. PhD dissertation, North Carolina State University, Raleigh.

38. E. J. Wyers, M. B. Steer, C. T. Kelley, and P. D. Franzon. 2013. A Bounded and Discretized Nelder-Mead Algorithm Suitable for RFIC Calibration. *IEEE Trans. Circuits Syst. I Reg. Papers* 60(7): 1787–1799.

39. J. E. Dennis Jr. and D. J. Woods, Optimization on Microcomputers: The Nelder-Mead Simplex Algorithm," in *New Computing Environments: Microcomputers in Large-Scale Computing*, A. Wouk, Ed. Philadelphia, PA: SIAM, 1987, pp. 116–122.

40. W. R. Klingman and D. M. Himmelblau. 1964. Nonlinear Programming with the Aid of a Multiple-Gradient Summation Technique. *J. ACM* 11(4): 400–415.

41. E. J. Wyers, M. A. Morton, T. C. L. G. Sollner, C. T. Kelley, and P. D. Franzon. 2016. A Generally Applicable Calibration Algorithm for Digitally Reconfigurable Self-Healing RFICs. *IEEE Trans. Very Large Scale Integr. (VLSI) Syst.* 24(3): 1151–1164.

42. Y.-S. Shu, J. Kamiishi, K. Tomioka, K. Hamashita, and B.-S. Song. 2010. LMS-Based Noise Leakage Calibration of Cascaded Continuous-Time $\Delta\Sigma$ Modulators. *IEEE J. Solid-State Circuits* 45(2): 368–379.

43. S. Saleem and C. Vogel. 2011. Adaptive Blind Background Calibration of Polynomial-Represented Frequency Response Mismatches in a Two-Channel Time-Interleaved ADC. *IEEE Trans. Circuits Syst. I Reg. Papers* 58(6): 1300–1310.

44. A. Swaminathan, K. Wang, and I. Galton. 2007. A Wide-Bandwidth 2.4 GHz ISM Band Fractional-N PLL with Adaptive Phase Noise Cancellation. *IEEE J. Solid-State Circuits* 42(12): 2639–2650.

45. L. Der and B. Razavi. 2003. A 2-GHz CMOS Image-Reject Receiver with LMS Calibration. *IEEE J. Solid-State Circuits* 38(2): 167–175.

46. W. Qi, E. J. Wyers, Z. Yan, and P. D. Franzon. 2014. An Automated Test Infrastructure for NBTI Effect Investigation and Calibration in Digital Integrated Circuits. In *Proceedings of IEEE Silicon Errors in Logic – System Effects (SELSE)*, pp. 14–18, Stanford University, CA, April 1–2, 2014.

47. M. S. Bazaraa, H. D. Sherali, and C. M. Shetty. 2006. *Nonlinear Programming: Theory and Algorithms*. Hoboken, NJ: Wiley.

48. J. C. Lagarias, J. A. Reeds, M. H. Wright, and P. E. Wright. 1998. Convergence Properties of the Nelder-Mead Simplex Method in Low Dimensions. *SIAM J. Optim.* 9(1): 112–147.

49. J. E. Dennis Jr. and V. Torczon. 1991. Direct Search Methods on Parallel Machines. *SIAM J. Optim.* 1(4): 448–474.

50. K. I. M. McKinnon. 1998. Convergence of the Nelder-Mead Simplex Method to a Nonstationary Point. *SIAM J. Optim.* 9(1): 148–158.

51. V. Torczon and M. W. Trosset. 1998. From Evolutionary Operation to Parallel Direct Search: Pattern Search Algorithms for Numerical Optimization. In *Proceedings of the Symposium Interface: Computing Science and Statistics*, vol. 29, pp. 396–401. Houston, TE, U May 14–17, 1997.

52. D. L. Keefer. 1973. Simpat: Self-Bounding Direct Search Method for Optimization. *Ind. Eng. Chem. Process Des. Dev.* 12(1): 92–99.

53. Y.-L. Hour, M. Yang, R.-M. Zhao, and H. Lian. 2010. Initialization for Synchronous Sequential Circuits Based on Chaotic Particle Swarm Optimization. In *Proceedings of the International Conference on Machine Learning and Computing (ICMLC)*, Bangalore, February 9–11, pp. 1566–1571.

54. R. Jenssen, D. Erdogmus, K. K. Hild II, J. C. Principe, and T. Eltoft. 2005. Optimizing the Cauchy-Schwarz PDF Distance for Information Theoretic, Non-Parametric Clustering. In *Energy Minimization Methods in Computer Vision and Pattern Recognition*. Lecture Notes in Computer Science 3757. Berlin: Springer, pp. 34–45.

55. R. Bramley and A. Sameh. 1992. Row Projection Methods for Large Nonsymmetric Linear Systems. *SIAM J. Sci. Statist. Comput.* 13(1): 168–193.

56. F. I. Bashir, A. A. Khokhar, and D. Schonfeld. 2006. View-Invariant Motion Trajectory-Based Activity Classification and Recognition. *Multimedia Syst.* 12(1): 45–54.

57. R. H. Bielschowsky and F. A. M. Gomes. 2008. Dynamic Control of Infeasibility in Equality Constrained Optimization. *SIAM J. Optim.* 19(3): 1299–1325.

58. D. Tzikas and A. Likas. 2010. An Incremental Bayesian Approach for Training Multilayer Perceptrons. In *Artificial Neural Networks – ICANN 2010*, Thessaloniki, Greece, September 15–18. Lecture Notes in Computer Science 6352, pp. 87–96.

59. S. Jeon, Y.-J. Wang, H. Wang, et al. 2008. A Scalable 6-to-18 GHz Concurrent Dual-Band Quad-Beam Phased-Array Receiver in CMOS. *IEEE J. Solid-State Circuits* 43(12): 2660–2673.

60. M. Brownlee, E. J. Wyers, K. Mayaram, and U.-K. Moon. 2002. PLL Design in SOI. In CDADIC Report, https://scholar.google.com/scholar?cluster=517120495 3264227553&hl=en&as_sdt=5,34&sciodt=0,34.

61. H. Wang and A. Hajimiri. 2007. A Wideband CMOS Linear Digital Phase Rotator. In *Proceedings of IEEE Custom Integrated Circuits Conference (CICC)*, San Jose, CA, September 16–19, pp. 671–674.

62. F. Bohn, K. Dasgupta, and A. Hajimiri. 2011. Closed-Loop Spurious Tone Reduction for Self-Healing Frequency Synthesizers. In *Proceedings of IEEE Radio Frequency Integrated Circuits Symposium (RFIC)*, Baltimore, MD, June 5–7, pp. 1–4.

63. M. Brownlee, E. J. Wyers, K. Mayaram, and U.-K. Moon. 2003. Radiation Hard PLL Design Tolerant to Noise and Process Variations. In CDADIC Report, https://scholar.google.com/scholar?cluster=1955362638137547388&hl=en &as_sdt=5,34&sciodt=0,34.

64. K. Bernstein, D. J. Frank, A. E. Gattiker, et al. 2006. High-Performance CMOS Variability in the 65-nm Regime and Beyond. *IBM J. Res. Dev.* 50(4–5): 433–449.

65. C. Hu. 1983. Hot-Electron Effects in MOSFETs. In *Proceedings of International Electron Devices Meeting (IEDM)*, Washington, DC, December 5–7, vol. 29, pp. 176–181.

66. C. Audet and J. E. Dennis Jr. 2001. Pattern Search Algorithms for Mixed Variable Programming. *SIAM J. Optim.* 11(3): 573–594.

67. J. Kennedy and R. C. Eberhart. 1995. Particle Swarm Optimization. In *Proceedings of IEEE International Conference on Neural Networks (ICNN)*, Perth, WA, November 27–December 1, pp. 1942–1948.

68. D. E. Finkel. 2005. Global Optimization with the DIRECT Algorithm. PhD dissertation, North Carolina State University, Raleigh.

69. S. Kirkpatrick, C. D. Gelatt Jr., and M. P. Vecchi. 1983. Optimization by Simulated Annealing. *Science* 220(4598): 671–680.

70. M. Mohri, A. Rostamizadeh, and A. Talwalkar. 2012. *Foundations of Machine Learning.* Cambridge, MA: MIT Press.

71. P. Chen, B. M. Merrick, and T. J. Brazil. 2015. Bayesian Optimization for Broadband High-Efficiency Power Amplifier Designs. *IEEE Trans. Microw. Theory Tech.* 63(12): 4263–4272.

72. C. M. Bishop. 2006. *Pattern Recognition and Machine Learning.* New York: Springer.

73. A. Forrester, A. Sobester, and A. Keane. 2008. *Engineering Design via Surrogate Modeling: A Practical Guide.* Hoboken, NJ: Wiley.

74. M. B. Yelten, T. Zhu, S. Koziel, P. D. Franzon, and M. B. Steer. 2012. Demystifying Surrogate Modeling for Circuits and Systems. *IEEE Circuits Syst. Mag.* 12(1): 45–63.

75. S. Sun, F. Wang, S. Yaldiz, et al. 2014. Indirect Performance Sensing for On-Chip Self-Healing of Analog and RF Circuits. *IEEE Trans. Circuits Syst. I Reg. Papers* 61(8): 2243–2252.

76. F. Wang, M. Zaheer, X. Li, J.-O. Plouchart, and A. Valdes-Garcia. 2015. Co-Learning Bayesian Model Fusion: Efficient Performance Modeling of Analog and Mixed-Signal Circuits Using Side Information. In *Proceedings of IEEE/ACM International Conference on Computer-Aided Design*, Austin, TX, November 2–6, pp. 575–582.

10

Role of Diffusional Interfacial Sliding during Temperature Cycling and Electromigration-Induced Motion of Copper Through Silicon Via

Lutz Meinshausen, Ming Liu, Indranath Dutta, Tae-Kyu Lee and Li Li

CONTENTS

ABSTRACT Stacked integrated circuits with copper-filled through silicon vias (TSVs) are a common component of three-dimensional integration concepts in microelectronics. The interfaces of the resulting Cu/Si composite material are affected by diffusional interfacial sliding which leads to TSV intrusion or protrusion. The resulting differential strain of the Cu TSV and the Si matrix affects the performance of the transistor or the reliability of interconnects close to the TSV. This may limit the use of stacked integrated circuits under harsh environment conditions.

The influence of temperature cycling (TC) and electromigration on the motion of a fully developed 10 μm Cu TSV is investigated. The experimental result shows a dwell time–dependent saturation of the TSV protrusion or intrusion during TC. Continuous TSV motion was observed during the electromigration test. The relevance of diffusional interfacial sliding for the TSV motion was shown with a three-dimensional finite-element model by incorporating interfaces with diffusion-controlled creep properties.

10.1 Introduction

Chip-on-chip (CoC) means to stack multiple chips vertically on top of each other. It enables the placement of integrated circuits and analogue devices like sensors in one package. Reducing the footprint and the R-C delay, CoC is of special interest for autonomous machines with an advanced sensor system to handle complex environments, like autonomous cars in heavy traffic. In addition, the integrated sensor systems have to have a long life cycle (20–30 years) under harsh environment conditions. CoC structures are based on Cu-filled through silicon vias (TSVs) that vertically connect the single chips with each other. Under harsh environment conditions, the TSVs are subjected to thermomechanical cycling during service. The difference in thermal expansion between Cu (16.4 ppm/K) and Si (2.5 ppm/K) leads to compressive and tensile stresses in the Cu and the surrounding Si matrix. Temperature cycling (TC) shows that the resulting relative expansion and shrinkage of the TSV compared with the Si matrix leads to cracks at the interfaces between the TSV and the lower and upper metal lines and at the TSV sidewalls [1,2]. In particular, cracks at the interfaces between the TSV and the metal lines increase the resistance of the TSV connection and alter the purpose of the three-dimensional packaging concept [1]. Furthermore, stresses in the Si matrix degrade the performance of front-end structures like proximate transistors [3].

In principle, a chip with TSVs is a composite with metal–nonmetal interfaces between Cu and Si. Although either a perfectly bonded or debonded interface was assumed in most analytical models, cycling experiments on graphite fibre–reinforced Al composites indicated differential matrix and fibre strains (i.e. a relative translation between the fibre and the matrix, non-isostrain) without interfacial fracture due to the diffusional accommodated recovery process and viscous drag in the highly dislocated interface region [4–6]. Further push-down tests with a tungsten plunger on quartz and Ni fibres in a Pb matrix could identify diffusional interfacial creep of well-bonded interfaces as the main mechanism of the fibre motion under shear stresses [7]. In [7], the experimental results were rationalised by an analytical one-dimensional model of the diffusional creep-driven interfacial sliding:

$$\dot{U} = \frac{8\Omega\delta_i D_i}{k_B T h^2} \cdot \left[\tau_i + 2\pi^3 \left(\frac{h}{\lambda} \right)^3 \sigma_n \right] \tag{10.1}$$

The diffusional interfacial sliding rate ($\dot{U}$) depends on the far-field interfacial shear stress (τ_i) and the interfacial diffusivity ($\delta_i D_i$). The sliding rate further depends on the shape of the interface, like its roughness (h) and the interfacial periodicity (λ). Normal (radial) stresses at the interface (σ_n), appearing due to the shear stresses along the periodic interface and the residual hydrostatic stresses in the TSV, influence the TSV motion as well. Tensile hydrostatic stresses lead to positive radial stresses ($\sigma_n > 0$), which accelerate interfacial sliding, while compressive hydrostatic stresses lead to negative radial stresses ($\sigma_n < 0$), which slow down interfacial sliding.

Investigations on back-end structures of integrated circuits have shown the presence of diffusional interfacial sliding along Cu/Si interfaces [8]. The initial stress in the Cu after electroplating is tensile [9–11]. The tensile stresses in Cu appear due to thermal expansion mismatch, self-annealing [12–15] and segregation of impurities during plating [9].

Observations of interfacial sliding along metal noninterfaces under the influence of a current flow were observed if the electromigration (EM)-induced material flow at the interfaces was much greater than the EM flow at the grain boundaries [16]. The EM-induced interfacial sliding rate depends on the effective charge of the moving ion (Z^*), the electronic charge (e) and the applied electric field (E).

$$\dot{U} = \frac{4\delta_i D_i}{k_B T h} Z^* e E \tag{10.2}$$

In [17], first investigations on the interfacial sliding of Cu TSVs in Si substrates were performed. The plastic deformation of the TSV "TSV-Pumping" is indicated by a bulging Cu surface. In contrast the interfacial sliding of the TSVs leads to a plane Cu surface and forms a step between TSV and Si surface. Further, the reviled interfaces have a smooth surface without Cu residuals, which excludes the presence of TSV sliding after interfacial fracture. Both observations indicate the presence of diffusional interfacial sliding along the bonded Cu/Si interfaces. The test results for annealed and nonannealed TSVs also indicated that an initial tensile stress leads to TSV intrusion during TC, while lower initial stresses after the annealing lead to TSV protrusion. The results also showed an influence of EM on the sliding rate. In particular, the TSV motion during TC slowed down after numerous cycles, while the EM-induced siding rate kept constant. As shown in Figure 10.1, the shear stress and the EM-induced interfacial sliding can be incorporated in one analytical model:

$$\dot{U} = \frac{8\Omega\delta_i D_i}{k_B T h^2} \cdot \left[\tau_i + 2\pi^3 \left(\frac{h}{\lambda} \right)^3 \sigma_n \right] + \frac{4\delta_i D_i}{k_B T h} Z^* e E \tag{10.3}$$

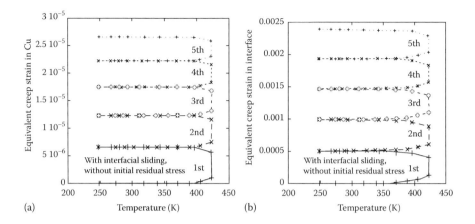

FIGURE 10.1
Schematic of the layout of the EM test structure (a). Interfacial sliding along a periodic interface, due to EM (v_{EM}) as a result of the applied current and shear stresses due to the Coefficient of thermal expansion (CTE) mismatch between Cu and Si (v_τ). In addition, the TSV motion and its thermal expansion cause normal stresses (σ_n), slowing down the sliding process.

First, investigations on diffusional interfacial sliding of Cu TSVs were done on prototypes without interfacial engineering. Hence, weak interfaces were one possible reason for the presence of interfacial sliding during TC or EM tests. Against this background, the following investigations were performed on fully developed samples with adhesion and buffer layers between the TSVs and the Si substrate. The TSV diameter was 10 μm and the wafer thickness 100 μm. Compared with previous investigations, the new samples should show a lower interfacial diffusivity, which would slow down diffusional interfacial sliding. Nevertheless, they also should show reduced radial stresses compared with larger TSVs without buffer layers. In the case of compressive or tensile radial stresses, the presence of a buffer layer might either accelerate or slow down interfacial sliding. To validate these assumptions, TC tests with a slow (0.1 K/s) and a fast (15 K/s) cycling rate were performed. The chosen temperature range was from –25°C to 150°C for the slow cycling and from 25°C to 150°C for the fast TC tests. Further EM tests were performed at 120°C and 170°C with an applied current density of 5×10^5 A/cm² in Cu TSVs.

One-dimensional analytical models for the description of the TSV under the influence of shear stresses are already available; nevertheless, TSV stresses are three-dimensional in nature with distinct near-surface characteristics. Near-surface stresses are important in leading to Cu fibre extrusion or intrusion relative to the Si surface. Hence, the intrusion or protrusion of Cu fibre relative to the Si surface during thermal cycling was studied with an axisymmetric finite-element (FE) model, incorporating interfacial sliding, multiple creep mechanisms, plasticity and manufacturing stress.

The aim of the FE simulations is a better understanding of interfacial sliding and the effects of initial stress and adjustable parameters of thermal cycling.

10.2 Modelling

10.2.1 Material Behaviour at the Interfaces

Cu fibre and Si matrix are assumed to be thermoelastic plastic creep and thermoelastic, respectively. The interface has the same material properties as Cu fibre expect for the creep behaviour. The cases with and without an interfacial sliding region have to be compared. For the case without a viscous interface, the interface region has the same creep property as Cu fibre. Isotropic behaviour is assumed for Cu due to its overall random orientation in TSVs [17] in spite of the anisotropic nature of Cu [27,28]. The change in microstructure during thermal cycling is little [17] and not considered, although grain growth was observed under isothermal holding conditions [29]. Additive strain rate decomposition, in which plastic and viscous networks for Cu are in series, is used, and thus the total strains ($\bar{\varepsilon}_f$ and $\bar{\varepsilon}_m$ for fibre and matrix, respectively) are given by [4]

$$\bar{\varepsilon}_f = \bar{\varepsilon}_f^{th} + \bar{\varepsilon}_f^{el} + \bar{\varepsilon}_f^{pl} + \bar{\varepsilon}_f^{cr} \tag{10.4}$$

$$\bar{\varepsilon}_m = \bar{\varepsilon}_m^{th} + \bar{\varepsilon}_m^{el} \tag{10.5}$$

where superscripts th, el, pl and cr represent thermal, elastic, plastic and creep components, and subscripts f and m represent fibre and matrix, respectively. Elastic strain, $\bar{\varepsilon}^{el}$, and thermal strain, $\bar{\varepsilon}^{th}$, are given by

$$\bar{\varepsilon}^{el} = \sigma/E \tag{10.6}$$

$$\bar{\varepsilon}^{th} = \alpha \Delta T \tag{10.7}$$

where E is elastic modulus and α represents coefficient of thermal expansion. Thermal expansion coefficients of Cu and Si are $17 \times 10^{-6}/K$ and $3 \times 10^{-6}/K$, respectively [24]. Poisson's ratios are 0.33 for Cu [25] and 0.28 for Si [26], respectively. The elastic modulus of Si is 130 GPa [24], and the elastic modulus of Cu is assumed to decrease linearly with temperature [24]:

$$E = 115\left[1 - (T - 300)/T_m\right] \tag{10.8}$$

The plasticity of Cu is considered, since the importance of plastic flow of the Cu via is well known [24]. Yield stress, σ_y, is dependent on temperature:

$$\sigma_y = 250\sqrt{\exp\left[-10(T-296)/1000\right]} \tag{10.9}$$

A linear strain hardening is used, and the linear hardening plastic modulus, E_t, is dependent on temperature [27]:

$$E_t = 1000\left(1-(T-300)/T_{\mathrm{m}}\right) \tag{10.10}$$

where $T_M = 1356$ K is the melting point of Cu.

The dominant creep mechanism operative in the Cu via changes continually during thermal cycling, and thus unified creep laws are necessary and imperative to adequately capture strain responses when temperature and manufacturing stress vary significantly [4,27,28]. The unified steady-state creep laws for Cu consist of Coble creep, $\dot{\gamma}_1$, and dislocation creep, $\dot{\gamma}_2$, driven by both lattice and core diffusion [29]:

$$\dot{\gamma}_f = \dot{\gamma}_1 + \dot{\gamma}_2 \tag{10.11}$$

where [34]

$$\dot{\gamma}_1 = 42 \frac{\tau\Omega}{kT} \frac{\pi\delta D_b}{d^3}, \dot{\gamma}_2 = A_2 D_{\mathrm{eff}} \frac{\mu b}{kT}(\tau/\mu)^n \tag{10.12}$$

and [34]

$$D_{\mathrm{eff}} = D_v\left[1+\frac{10a_c}{b^2}(\tau/\mu)^2 \frac{D_c}{D_v}\right], D_v = D_{0v}\exp\left(-\frac{Q_v}{RT}\right) \tag{10.13}$$

$$\delta D_{gb} = \delta D_{0gb}\exp\left(-\frac{Q_{gb}}{RT}\right), a_c D_c = a_c D_{0c}\exp\left(-\frac{Q_c}{RT}\right) \tag{10.14}$$

$$\mu = \mu_0\left(1+\frac{T-300}{T_M}\frac{T_M}{\mu_0}\frac{d\mu}{dT}\right), \frac{T_M}{\mu_0}\frac{d\mu}{dT} = -0.54 \tag{10.15}$$

where μ is the shear modulus and is dependent on temperature ($\mu_0 = 42.1$ GPa at 300 K); d is the grain size of 1 μm, which was found in Cu-filled TSVs under an annealing temperature of 400°C [23]; $R = 8.314$ J/K/mol is the gas constant; T is the absolute temperature; $\Omega = 1.18 \times 10^{-29}$ is the atomic volume of Cu; $b = 2.56 \times 10^{-10}$ is Berger's vector; $D_{0v} = 2\times 10^{-5}$ and $Q_v = 197$ kJ/mol are the preexponential factor and activation energy for lattice diffusion, respectively; $\delta D_{0gb} = 5 \times 10^{-15}$ m³/s and $Q_{gb} = 104$ kJ/mol are the preexponential factor and activation energy for grain boundary diffusion, respectively; $a_c D_{0c} = 1 \times 10^{-24}$ m^4/s and $Q_c = 117$ kJ/mol are the preexponential factor and activation energy for core diffusion, respectively; $n = 4$ is the dimensionless exponent for power-law creep; and $A_2 = A(\sqrt{3})^{n+1}$, with $A = 7.4 \times 10^5$ being a dimensionless constant for power-law creep.

Interfacial sliding is accounted for via diffusion-controlled creep mechanism without breaking interfacial bonds [4,30,31], since the interface was found to be a high diffusivity path and slide via diffusion-controlled creep [32]. Diffusion-accommodated interfacial sliding does not result in interfacial fracture [24], and a continuum model is adequate to describe interfacial sliding [32]. Interfacial sliding induced by the interfacial shear stress is a diffusion-controlled phenomenon. The interface is assumed to be microscopically periodic with a width of h and periodicity of λ [30,32]. The strain rate, $\dot{\gamma}_i$, due to interfacial sliding can be represented as [32–34]

$$\dot{\gamma}_i = A_i \tau_i \tag{10.16}$$

where τ_i is the shear stress acting on the interface, the subscript i indicates interface, and

$$A_i = \frac{8\delta_i D_{0i}\Omega}{kTh^3} \exp\left(-\frac{Q_i}{RT}\right) \tag{10.17}$$

where Q_i and $\delta_i D_{0i}$ are activation energy and preexponential parameter for interfacial diffusion, respectively. The interfacial width h was taken to be 0.1 μm [4,32], and $Q_i = 55$ kJ/mol, $\delta_i D_{0i} = 10^{-4} \delta D_{0gb}$ [24]. Those parameters ensure that the interface serves as a rapid diffusion path.

In FE simulation, the relation between shear strain rate, $\dot{\gamma}$, and shear stress, τ, is converted to the relation between the equivalent strain rate, $\dot{\varepsilon}$, and equivalent von Mises stress, $\bar{\sigma}$:

$$\tau = \frac{\bar{\sigma}}{\sqrt{3}}, \dot{\gamma} = \sqrt{3}\dot{\varepsilon} \tag{10.18}$$

10.2.2 Material Behaviour at the Interfaces Simulation of the Manufacturing Process and Thermal Cycling

Initial stress induced during the manufacturing process before the start of thermal cycling was found to play a significant role in relative displacement of the Cu pillar with respect to the substrate [24]. Therefore, residual stress before thermal cycling, which is inevitable in multicomponent structures, is considered, although details of TSV fabrication processes are not known [19]. For the case with the manufacturing process occurring before thermal cycling, the initially stress-free state is assumed to be at 100°C, and the following conceptual thermal excursion is applied: (1) cooling to 20°C at a very large cooling rate so that creep deformation is prevented; (2) heating to 400°C [19], which is the temperature of the chemical vapour deposition process for depositing an insulation layer of oxide on the wall of TSVs [35], at a very large heating rate ignoring creep mechanism; (3) holding time of 10 min at 400°C, which is the annealing temperature for etch-stop deposition [18,19]; (4) cooling to 20°C at 10°C/min; (5) holding for 24 h at 20°C. The thermal cycling is from –25°C to 150°C with a slope of 0.1°C/s and dwell time of 10 min. The annealing process and room temperature stress relaxation [36,37] can reduce residual stress inside the vias, and the measured stresses in the as-received sample (no thermal cycling and stored at room temperature for more than 9 months after its fabrication) can be very low [19]. Therefore, the case without residual stress (i.e. free stress state at room temperature, 20°C, before thermal cycling) is used for comparison.

10.3 Experiments and FE Analysis

10.3.1 Experiments

The test samples were 100 μm thick Si dies with Cu TSV arrays. As shown in Figure 10.2, the TSVs are 10 μm in diameter and the interfaces are covered with a buffer layer. The pitch between the TSVs varies between 15 and 75 μm.

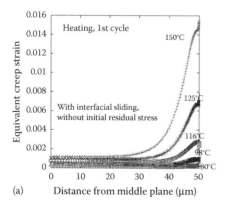 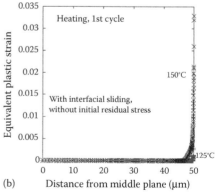

FIGURE 10.2

(a,b) SEM pictures showing buffer layers and diffusion barriers around the TSV.

The back- and front-end structures were removed by polishing. Finally, the TSVs were exposed at both surfaces of the die. The exposed TSVs were finally polished with 0.05 μm colloidal silica. The colloidal silica removed the Si faster than the Cu, leading to an initial protrusion of the TSVs.

The differential length (Δl) between the Cu TSV (L_{TSV}) and the Si (L_{Si}) was measured with scanning white light interferometry (SWLI). Due to the limited lateral resolution of swli, the exposed Cu/Si interfaces were observed by scanning electron microscopy (SEM) to identify possible signs of interfacial sliding or interfacial fracture. During the observations, the samples were tilted by 55°. The surface profiles were measured along three lines through the centre of the TSV. For every data point, four TSVs were observed.

$$\Delta l = \overline{L_{TSV}} - \overline{L_{Si}} \tag{10.19}$$

TC was performed at the following conditions: (1) between –25°C and 150°C with heating and cooling rates of 0.1 K/s and a dwell time of 10 min (slow cycling) and (2) between 25°C and 150°C with heating and cooling rates of 15°C/s and a dwell time of 2 min (rapid cycling). The EM test temperatures were 120°C (12–120 h) and 170°C (24–72 h). The current density was 5×10^5 A/cm² in the Cu TSV. The EM tests were performed in a vacuum furnace to avoid side effects by Cu oxidation.

10.3.2 Material Behaviour at the Interfaces FE Simulations

The radius and height of the modelled Cu via are 5 and 100 μm, respectively. The radial dimension of the full model is 505 μm, under which conditions Si can be regarded to be infinitely large. In order to study the interfacial sliding [20,21], an ultrathin interfacial layer between Cu and Si is considered. The width of the interface is 0.1 μm, which is a reasonable value, since the width of the interface zone was found to be 0.3–0.45 μm for graphite fibre–reinforced Al composite [4]. An axisymmetric FE model is built due to the symmetry of the problem, with the axial coordinate z being zero at the middle of the Cu fibre length, and the radial coordinate r being zero at the middle of the Cu fibre diameter. Only the top half of the structure is modelled, since the shear stress–driven sliding occurs symmetrically at both ends of the via, since the induced stress is symmetrical across the middle of the via [22]. Therefore, the axial dimension of the modelling structure is half of the height of the Cu via, which is 50 μm. Zero axial displacement is set at the middle plane (i.e. $z = 0$), and zero radial displacement is set at the axial axis (i.e. $r = 0$). Radial displacements are kept the same for the nodes on the outer surface, in order to mimic an infinitely large matrix.

10.4 TSV Motion during Slow Thermal Cycling

10.4.1 Experimental Results

Before the TC experiments were started, SWLI measurement after polishing showed an initial TSV protrusion of 25 nm on both sides of the die. The TC was performed in a temperature range between –25°C and 150°C with dwell times of 10 min. The slope was 0.1 K/s. After TC, the difference in the TSV protrusion compared with initial results indicates the TSV motion during cycling. The TSV position relative to the Si was measured with SWLI after 5, 15, 15, 50 and 100 cycles. In addition, the revealed interfaces were observed with SEM to differ between diffusional interfacial sliding from plastic deformation or sliding after interfacial fracture.

The average TSV position during TC is shown in Figure 10.3. During the first five cycles, the TSV shrank by 70 nm and was finally intruded by 10 nm. Previous investigations on TSVs show the formation of voids in the TSV during high temperature cycles which appear during Cu annealing or during further fabrication steps [2]. The samples certainly passed high temperature cycles during the fabrication of the back-end structures on their top and bottom sides. Furthermore, the temperature treatment of Cu increases the average grain size and reduces its yield strength, hardness and elastic modulus [37]. During the following cycles, the TSV starts to protrude until the differential strain between Cu and Si equalises for 150°C and the protrusion saturates.

Figures 10.4 and 10.5 show the SWLI measurements as well as SEM pictures of the revealed interfaces. The case of intrusion, as well as for protrusion, of the TSVs revealed that Si and Cu surfaces do not show any sign of interfacial fracture, like residual Cu at the revealed Si surface. Neither the SEM pictures nor the SWLI data show concave or convex bulging of the intruded or protruded TSVs. Hence, the clear steps between the Cu TSV and the Si matrix, and the smooth interfaces indicate the presence of diffusional interfacial sliding during the slow TC tests.

10.4.2 Material Behaviour at the Interfaces FE Simulations

Figure 10.6 shows the distribution of axial displacement at 20°C after 200 thermal cycles for the cases with interfacial sliding. The transverse plastic/creep deformation in compression makes the via expand axially [22], and thus protrusion of Cu relative to Si is expected. The results with and without residual stress are compared. The critical points, whose axial displacements are monitored, are highlighted. Boundary conditions are also displayed. Axial displacement is quite nonuniform across the interface surface due to the interfacial sliding. Residual stress has a significant effect on the deformation of the interface. Displacement on the Si surface is small and almost

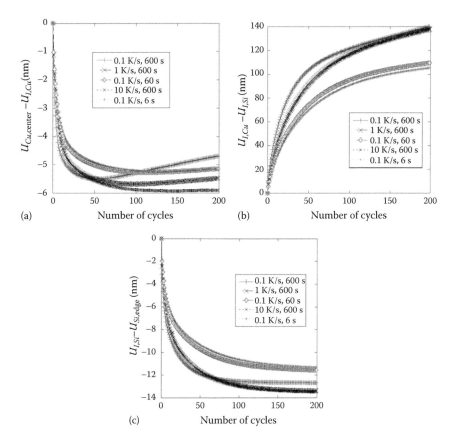

FIGURE 10.3
(a–c) Average TSV position during slow TC. During the first cycles, the TSV intrudes. Afterwards, TSV protrusion can be observed. The TSV protrusion slows down after several cycles.

uniform due to its purely thermoelastic property. For the case without residual stress, axial displacements are quite uniform on the Cu surface, and the top point is located at the interface end, while for the case with residual stress, the manufacturing process induces protrusion of the Cu via out of the Si matrix before thermal cycling, and axial displacement is not uniform on the Cu surface, with the top point being located at the surface centre. For the case with residual stress, a small intrusion of Cu relative to Si is found near the interface, which is consistent with an experimental finding that the top region of the Cu TSV was found in compression after the manufacturing process [18]. The bow-like stress profile, which was also reported in a previous FE study [38], is attributed to the varying constraint: less constraint experiences low stress at the fibre end.

FIGURE 10.4
(a) After the initial several cycles, TSV intrusion was observable by SEM. (b) SWLI shows a TSV intrusion by a few nanometres. (c) A closer view on the interfaces does not show any remains of Cu, indicating interfacial sliding rather than interfacial fracture.

FIGURE 10.5
TSV protrusion after 100 cycles. The SEM (a) and SWLI (b) indicate a TSV protrusion of more than 100 nm. (c) Smooth interface indicates interfacial sliding again.

Figure 10.7 compares the relative axial displacements for the cases with interfacial sliding. $U_{Cu,center} - U_{I,Cu}$ and $U_{I,Si} - U_{Si,edge}$ indicate the surface profiles of the Cu via and Si matrix, respectively; $U_{I,Cu} - U_{I,Si}$ represents the interfacial sliding. During thermal cycling, the surface profiles of the Cu via and Si matrix remain almost unchanged after the initial several cycles, while interfacial sliding varies prominently, with the changing rate decreasing

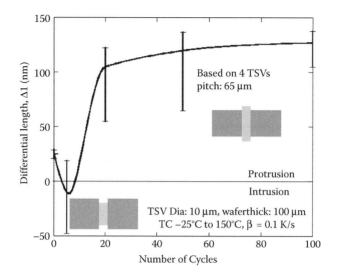

FIGURE 10.6
Axial displacement field (μm) at 20°C after 199 cycles for the cases with interfacial sliding: (a) without and (b) with residual stress before thermal cycling.

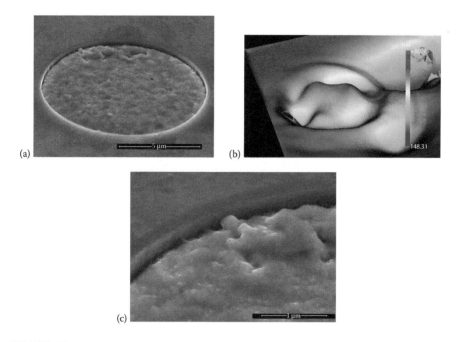

FIGURE 10.7
Relative axial displacements at 20°C after each cycle for the cases with interfacial sliding: (a) without and (b) with residual stress.

with the number of cycles. The creep rate decreases gradually, since the matrix Si is much stiffer and stronger without any creep [1]. Compressive longitudinal strain is induced in the Si matrix, and the magnitude of the axial strain in the Si matrix is much smaller than that for Cu fibre. Clearly, fibre and matrix are not in isostrain conditions, and Cu fibre protrudes relative to the Si matrix. It is the interface with diffusional sliding that accommodates the large differential strains between the matrix and the fibre in the absence of debonding or frictional sliding. Residual stress determines the surface profile and the direction of interfacial sliding: for the case without residual stress, interfacial sliding increases, resulting in an increase in protrusion; for the case with residual stress, interfacial sliding decreases from positive to negative values, resulting in a decrease of protrusion. For the case with residual stress, a steady state is reached after 60 thermal cycles, while for the case without residual stress, many more thermal cycles are needed in order to reach a steady state. The FE model utilises only steady-state creep, and thus the mechanical responses (i.e. creep rates) are expected to deviate somewhat from the actual material response [39]. Furthermore, the kinetics of interface creep is expected to be dependent on grain orientation, and the preexponential factor of grain boundary diffusion was found to change with grain orientation, considering a constant activation energy [40]. In reality, the interface is Cu/SiO_2 but not Cu/Si [40], and more reasonable parameters for interfacial creep are needed for better comparison with experimental results. A certain residual stress is expected in the samples used in the experiments, which explains the nonuniform surface profile of the Cu via, and tens of thermal cycles before a steady state is reached.

Volume average results are calculated according to

$$\bar{\chi} = \frac{\sum \chi_i V_i}{\sum V_i} \tag{10.20}$$

where χ_i is the variable of interest and V_i is the volume at the ith element in the deformed configuration.

Axial stress decreases during heating and increases during cooling (Figure 10.8). Although the stress hysteresis in one cycle is small (the hysteresis is more prominent during the first cycle, consistent with a previous finding for graphite fibre–reinforced Al composite [4]), since the dwell time at high temperature changes stress only a little, the stress keeps evolving with the number of thermal cycles increasing. The manufacturing process makes the Cu via protrude out of the Si matrix, and tensile longitudinal strain is induced in Cu fibre, consistent with the residual tensile axial stress. The decrease of protrusion for the case with residual stress is due to the shifting down of axial stress, and the increase of protrusion for the case without residual stress is due to the shifting up of axial stress. The evolution of volume average axial stress in the Cu region during thermal cycling is shown in

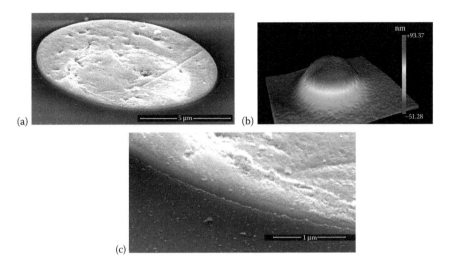

FIGURE 10.8
Volume average axial stress in Cu region during thermal cycling for the cases with interfacial sliding: (a) without residual stress and (b) with residual stress. The inset contour shows the contour of axial stress (MPa) at 20°C after 198 cycles.

Figure 10.8a and b: although the initial stress states are different depending on residual stresses, the stress states become similar after sufficient thermal cycling, and it is expected that the final stress state is independent of the stress state at the beginning of thermal cycling.

To show the effect of diffusional interfacial sliding on the overall TSV, an alternative model without interfacial was created. In this case, the interface region has the same creep behaviour as Cu. The relative axial displacement $U_{Cu,center} - U_{I,si}$ indicates the surface profile of the Cu via. As expected, there is little change of surface profile of the Si matrix during thermal cycling for the case without interfacial sliding (Figure 10.9). For the case with residual stress, the initial protrusion due to the manufacturing process decreases at a small rate during thermal cycling, while for the case without residual stress, Cu protrudes out of the Si matrix during thermal cycling at a small rate. It is expected that many more thermal cycles are needed before a steady state is reached. Although the role of the interface is a subject of confusion [4], it is clear that it is the interfacial sliding that results in the large relative displacement during thermal cycling, and without interfacial sliding, a larger number of thermal cycles is needed before stress reaches a saturation state and the TSV becomes stable.

Figure 10.10 shows the variation of equivalent creep strain during thermal cycling for the case without residual stress and with interfacial sliding. Creep deformation in the interface is much larger than that in the Cu region, since the interface is the fast diffusion path, and a significant creep deformation

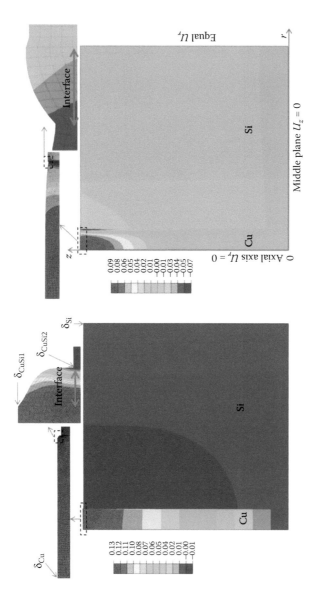

FIGURE 10.9
Relative axial displacements at 20°C after each cycle for the cases without interfacial sliding (relative displacements are zero for the case without residual stress, but nonzero for the case with residual stress).

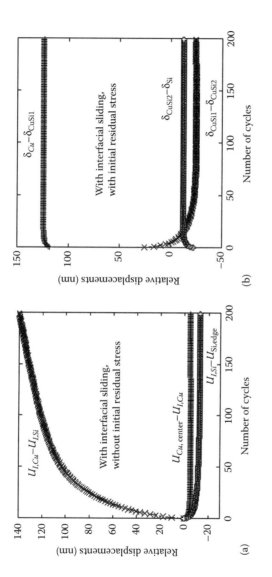

FIGURE 10.10
Volume average axial stress in Cu region during thermal cycling for the cases without interfacial sliding: (a) without residual stress and (b) with residual stress.

is induced during dwell time at high temperature. The increase of creep strain during one cycle decreases with thermal cycles, and it is expected that there would be little deformation due to the little creep deformation after sufficient cycles. The shifting up of axial stress accords with the increase of protrusion during thermal cycling for the case without the manufacturing process. The shifting down of axial stress is consistent with the decrease of protrusion during thermal cycling for the case with the manufacturing process.

In the model, the shear stress is zero near the middle plane, and significant interfacial shear stresses are induced at the surface–interface junction [22,24]. Singularities occur near the top surface in FE simulation [1]. After one cycle, the shear stress at the lowest temperature increases due to strain hardening, making the interfacial sliding increase for the case without residual stress and with interfacial sliding.

Due to the stress-free condition at the top surface, axial stress reaches zero at the fibre end. The axial stress is almost uniform near the middle plane, with the maximum value being at the middle plane. As expected, tensile axial stress is induced at low temperatures, since Cu contracts more than Si, and compressive axial stress is induced at high temperatures, since Cu of a larger thermal expansion coefficient expands more than Si. After each cycle (i.e. from –25°C to 125°C), axial stress near the top surface increases, consistent with the increase of protrusion during thermal cycling. During dwell time at the highest temperature, stress relieves via creep. Cu fibre creeps in compression for most of the time, consistent with a decrease of the relative axial displacements for the case without residual stress and with interfacial sliding

Figure 10.11 a and b shows the distribution of equivalent creep and plastic strains, respectively, during the first heating for the case without residual stress and with interfacial sliding. Creep strain accumulates during thermal cycling and occurs prominently at temperatures greater than 80°C. The region with the most constraint is located at the middle plane [18], and it can be approximated that the majority of Cu fibre, except for the top end, is in a hydrostatic stress state during thermal cycling. The maximum shear and von Mises stresses are located at the top end.

10.4.3 Material Behaviour at the Interfaces Discussion

The TC of the Cu TSV with a relative slow rate of 0.1 K/s has shown a protrusion of TSVs which saturates after numerous cycles. SEM pictures have not shown significant bulging of the TSV or residuals at the revealed interfaces. Instead, clear steps between the protruded Cu TSV and the Si matrix were observed. Hence, the experimental results indicate interfacial sliding as the main reason for TSV protrusion. Compared with the experimental results, the FE simulations incorporating a zone with diffusional interfacial sliding between Cu and Si show similar sliding rates and a saturation of the TSV

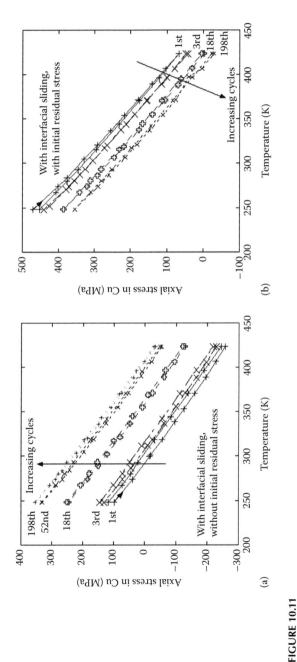

FIGURE 10.11
Volume average value of equivalent creep strain during thermal cycling for the case without residual stress and with interfacial sliding: (a) in Cu region and (b) in interface region.

protrusion, while FE models without a sliding interface did not show a sufficient TSV motion after 200 TCs. The strain rate due to diffusional sliding rapidly increases for temperatures above 80°C and stays at a high level during the dwell at 150°C. Hence, it can be concluded that diffusional interfacial sliding during the high-temperature phase of the TC is responsible for the TSV protrusion.

Like shown in previous investigations, the direction of the TSV motion depends on the initial stresses in the TSV [17]. The simulations show that the TSV protrusion goes along with an increasing tensile average volume stress in the Cu. In compliance with a saturation of the TSV protrusion, the compressive stresses in the Cu are reduced with every cycle, while larger tensile stresses are built up. The TSV protrusion during the TC experiments and the simulation results indicate that the initial stresses in the TSV samples were relatively low before testing.

10.5 TSV Motion during Fast Thermal Cycling

The FE simulation has shown that the TSV protrusion during the slow TC experiments mainly appeared due to diffusional interfacial sliding during the high-temperature phase ($T > 80°C$). Diffusional interfacial sliding is a time-dependent mechanism. Hence, a faster slope and shorter dwell time should reduce the TSV motion rate, if diffusional interfacial sliding is the main mechanism of TSV motion. Furthermore, the suppression of the diffusion-driven relaxation mechanism will increase the shear stress at the interfaces and may cause interfacial fracture. The TSV protrusion saturated after 20 cycles at a level of 45 nm (Figure 10.12). As expected, the protrusion of the TSV after the fast TC test is smaller than after an equivalent number of slow TCs. A possible reason for the reduced mobility of the TSV is a reduced influence of diffusional interfacial sliding and work hardening of the Cu.

The SWLI results for the fast cycling, as well as a closer observation of the revealed Cu surface with SEM, show no evidence of Si/Cu interfacial fracture (Figure 10.13). The smooth surface of the revealed interface indicates that the shear stresses during fast TC were still not large enough to cause interfacial fractures. Hence, diffusion interfacial sliding can appear along the bonded interfaces.

FE simulations with different heating and cooling rates, as well as different dwell times, were performed to evaluate the influence of the different TC test conditions on the TSV motion rate. The simulation shows the relative displacements under different heating rates and dwell times for the case without residual stress and with interfacial sliding (Figure 10.14). Different

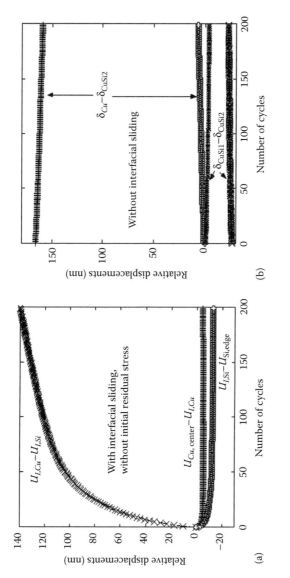

FIGURE 10.12
Mean position of the TSV for the fast Temperature cycling test (TCT). The TSV protrusions saturate after 15–20 cycles.

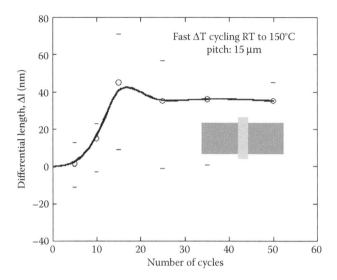

FIGURE 10.13
SWLI data (a) before and (b) after 50 fast temperature cycles, showing increased TSV protrusion of a few nanometres. (c, d) SEM picture of a protruded TSV after 50 temperature cycles, showing no interfacial fracture.

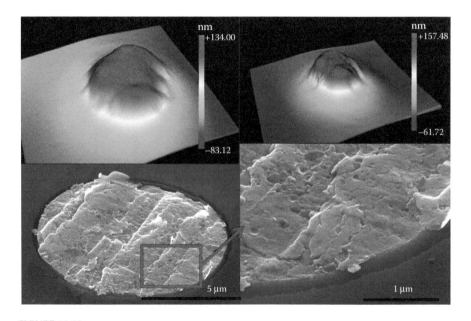

FIGURE 10.14
Effects of heating rate and dwell time on relative axial displacements at 20°C after each cycle for the case with interfacial sliding and without residual stress.

heating rates and dwell times have little effect on surface profiles of the Cu via and Si matrix during thermal cycling. The interfacial sliding at a sufficiently large number of cycles is independent of heating rates, and the dwell time has a prominent effect on the limit where a saturation effect of the interfacial sliding can be observed.

10.6 TSV Motion during EM Stress Tests

Interfacial sliding under EM conditions was observed at 120°C and 170°C. A current density of 5×10^5 A/cm^2 was applied to gain a sufficient driving force. SWLI data (Figure 10.15) and SEM pictures (Figure 10.16) were taken before and after the EM tests. The TSVs were observed to move along their interfaces in the direction of the electron flow.

During temperature cycling tests the TSV motion slows down over time. In contrast the EM-induced TSV motion was monotonically with time at a rate of 1.15 nm/h at 120°C and 3.18 nm/h at 170°C (Figure 10.17).

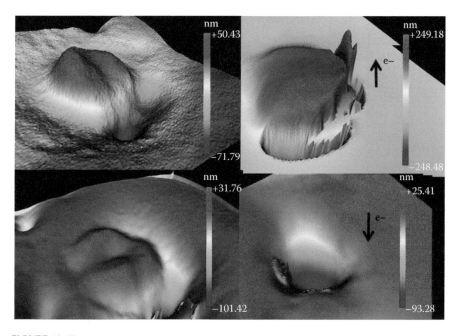

FIGURE 10.15
SWLI images showing that EM-driven interfacial siding leads to (a) TSV protrusion on one side and (b) intrusion on the opposite side. The EM test was performed for 62 h at 170°C.

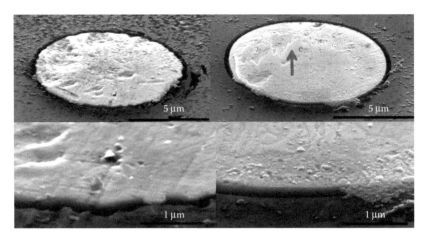

FIGURE 10.16
(a) Before the EM test, the TSV was nearly flush. (b) After 24 h EM testing at 170°C, the TSV protruded by more than 100 nm.

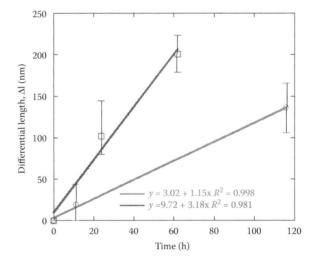

FIGURE 10.17
TSV motion during EM tests at two different temperatures. An increased test temperature leads to a faster TSV movement.

The interfacial sliding speed depends on the preexponential factor and activation energy for interfacial diffusion [10]. Equation 10.15 leads to an activation energy (E_A) of 0.31 eV for diffusional TSV sliding due to EM:

$$E_A = \ln\left(\frac{\dot{U}_2}{\dot{U}_1}\right) \cdot \frac{k_B T_1 \cdot T_2}{T_2 - T_1} \qquad (10.21)$$

An increased test temperature leads to compressive radial stresses at the Cu/Si interface. The radial stresses reduce the sliding rate of the TSV by increasing the threshold for EM-induced TSV motion. Hence, the calculated E_A is smaller than E_A for EM in the Cu/Si interface. The constant sliding rate of the TSVs under EM conditions makes EM an important long-term reliability issue in three-dimensional packages.

10.7 Conclusion

While TSV protrusion and intrusion due to diffusional interfacial sliding were already shown along direct Cu/Si interfaces in TSV prototypes with a diameter of 100 μm, it was not certain whether the same effect would also lead to TSV protrusion or intrusion of fully developed TSVs with 10 μm in diameter and stronger interfaces, including Cu seed layers and buffer materials around the TSVs.

Slow- and fast-rate TC tests showed an increasing TSV protrusion with every cycle, and SEM pictures of the revealed interface showed no signs of debonding or a significant bulging of the TSVs. To verify these experimental signs for diffusional interfacial sliding, FE simulations were also performed. In the FE model, diffusional controlled interfacial sliding is considered for a small region between Cu and Si. Unified creep laws for the Cu fibre are incorporated. The results show that interfacial sliding is the main cause of large relative displacements during TC, and tensile residual stresses lead to TSV intrusion, while stress-free TSVs protrude during TC. It is also found that the heating rate has little effect on the final deformation, while decreasing dwell times are an effective way to decrease interfacial sliding. The observation of a reduced TSV protrusion as a consequence of a reduced dwell time at maximum temperature of the TC tests is in compliance with the experimental results, which also have shown a reduced TSV protrusion during fast TC tests; furthermore, diffusional processes are time dependent in nature.

The EM tests show a unidirectional motion of the TSV. Like for the TC, the interfaces were not debonded during the tests. The motion speed of the TSV is exponentially temperature dependent as well. Hence, fully developed TSVs with relative strong interfaces are affected by EM-driven interfacial sliding along bonded interfaces. Like in previous EM tests on TSVs, the sliding rate is not reduced over time. Hence, EM can be an important long-term reliability issue of integrated circuits with TSVs.

Overall, the test results show that under harsh environment conditions, even fully integrated TSVs are prone to reliability risks due to interfacial sliding. The simulation results show the importance of diffusional interfacial sliding for the overall TSV motion. Under these conditions, TSVs can be an additional source of stress in neighbouring metal lines and transistors. Taking into account that temperature cycles or high operation temperatures cannot be avoided, the application of TSVs under harsh environment conditions would require further interfacial engineering. Beside a further increase of the activation energy for diffusional sliding along the metal–nonmetal interfaces, a rougher interface would also neutralise the stresses, causing the diffusional drift of the TSV.

Acknowledgements

This work was supported by grants from the National Science Foundation (NSF) (DMR-1309843), Cisco Systems and the Missile Defense Agency. The modelling work is supported in part by NSF through XSEDE resources provided by San Diego Supercomputer Center (SDSC) national allocation.

References

1. C. Okoro, J. W. Lau, F. Golshany, K. Hummler, Y. S. Obeng. A detailed failure analysis examination of the effect of thermal cycling on Cu TSV reliability. *IEEE Transactions on Electronic Devices* vol. 61, no. 1 (2014), pp. 15–22.
2. D. Zhang, K. Hummler, L. Smith, J. J. Q. Lu. Backside TSV protrusion induced by thermal shock and thermal cycling. In *IEEE 63rd Electronic Components & Technology Conference*, Las Vegas, May 2013, pp. 1407–1413.
3. H. Jao, Y. Y. Lin, W. Liao, B. Huang. The impact of through silicon via proximity on CMOS device. In *Microsystems, Packaging, Assembly and Circuits Technology Conference (IMPACT)*, Taipei, October 2012, pp. 43–45.
4. I. Dutta. Role of interfacial and matrix creep during thermal cycling of continuous fiber reinforced metal–matrix composites. *Acta Materialia* vol. 48 (2000), pp. 1055–1074.
5. S. Goto, M. Mclean. Role of interfaces in creep of fibre-reinforced metal-matrix composites – II short fibres. *Acta Metallurgica et Materialia* vol. 39, no. 2 (1991), pp. 153–164.
6. P. B. R. Nimmagadda, P. Sofronis. Creep strength of fiber and particulate composite materials: The effect of interface slip and diffusion. *Mechanics of Materials* vol. 23, no. 1 (1996), pp. 1–19.
7. J. V. Funn, I. Dutta. Creep behavior of interfaces in fiber reinforced metal-matrix composites. *Acta Materialia* vol. 47, no. 1 (1999), pp. 149–164.
8. I. Dutta, K. A. Peterson, C. Park. Modeling the interfacial sliding and film crawling in back-end structures of microelectronics devices. *IEEE Transactions on Components and Packaging Technologies* vol. 28, no. 3 (2005), pp. 397–407.
9. S. H. Brongersma, E. Kerr, I. Vervoort, et al. Grain growth, stress, and impurities in electroplated copper. *Journal of Materials Research* vol. 17, no. 3 (2002), pp. 582–589.

10. D. Gan, P. S. Ho, R. Huang, et al. Effect of a cap layer on morphological stability of a strained epitaxial film. *Journal of Applied Physics* vol. 97 (2005), p. 113537.

11. J. Auersperg, D. Vogel, E. Auerswald, et al. Nonlinear copper behavior of TSV and the cracking risks during BEoL-built-up for 3D-IC-integration. Presented at IEEE 13th EuroSimE, Lisbon, April 2012.

12. S. Lagrange, S. H. Brongersma, M. Judelewicz, et al. Self-annealing characterization of electroplated copper films. *Microelectronic Engineering* vol. 50 (2000), pp. 449–457.

13. C. Okoro, K. Vanstreels, R. Labie. Influence of annealing conditions on the mechanical and microstructural behavior of electroplated Cu-TSV. *Journal of Micromechanics and Microengineering* vol. 20 (2010), pp. 1–6.

14. H. Lee, S. S. Wong, S. D. Lopatin. Correlation of stress and texture evolution during self- and thermal annealing of electroplated Cu films. *Journal of Applied Physics* vol. 93 (2003), pp. 3796–3804.

15. K. Barmak, A. Gungor, C. Cabral Jr., J. M. E. Harper. Annealing behavior of Cu and dilute Cu-alloy films: Precipitation, grain growth, and resistivity. *Journal of Applied Physics* vol. 86, no. 5 (2005), pp. 2516–2525.

16. P. Kumar, I. Dutta. Effect of substrate surface on electromigration-induced sliding at hetero-interfaces. *Journal of Physics D* vol. 46 (2013), pp. 1–5.

17. P. Kumar, I. Dutta, M. S. Bakir. Interfacial effects during thermal cycling of Cu-filled through-silicon vias (TSV). *Journal of Electronic Materials* vol. 41, no. 2 (2012), pp. 322–335.

18. C. Okoro, L. E. Levine, R. Q. Xu, K. Hummler, Y. J. Obeng. Synchrotron-based measurement of the impact of thermal cycling on the evolution of stresses in Cu through-silicon vias. *Journal of Applied Physics* vol. 115 (2014), p. 243509.

19. J. Marro, C. Okoro, Y. Obeng, K. Richardson. Defect and microstructural evolution in thermally cycled Cu through-silicon vias. *Microelectronics Reliability* vol. 54, no. 11 (2014), pp. 2586–2593.

20. D. V. Zhmurkin, T. S. Gross, L. P. Buchwalter. Interfacial sliding in Cu/Ta/polyimide high density interconnects as a result of thermal cycling. *Journal of Electronic Materials* vol. 26, no. 7 (1997), pp. 791–797.

21. S. K. Ryu, K. H. Lu, X. F. Zhang, et al. Impact of near-surface thermal stresses on interfacial reliability of through-silicon vias for 3-D interconnects. *IEEE Transactions on Device and Materials Reliability* vol. 11, no. 1 (2011), pp. 35–43.

22. I. Dutta, P. Kumar, M. S. Bakir. Interface-related reliability challenges in 3-D interconnect systems with through-silicon vias. *JOM* vol. 63, no. 10 (2011), pp. 70–76.

23. A. Heryanto, W. N. Putra, A. Trigg, et al. Effect of copper TSV annealing on via protrusion for TSV wafer fabrication. *Journal of Electronic Materials* vol. 41 (2012), pp. 2533–2542.

24. P. Kumar, I. Dutta, M. S. Bakir. Interfacial effects during thermal cycling of Cu-filled through-silicon vias (TSV). *Journal of Electronic Materials* vol. 41 (2012), pp. 322–335.

25. D. J. Celentano, B. Guelorget, M. Francois, et al. Numerical simulation and experimental validation of the microindentation test applied to bulk elastoplastic materials. *Modelling and Simulation in Materials Science and Engineering* vol. 20, no. 4 (2012), pp. 1–12.

26. M. Liu, F. Q. Yang. Finite-element analysis of the indentation-induced delamination of Bi-layer structures. *Journal of Computational and Theoretical Nanoscience* vol. 9, no. 6 (2012), pp. 851–858.

27. I. Dutta, K. A. Peterson, C. Park, J. Vella. Modeling of interfacial sliding and film crawling in back-end structures of microelectronic devices. *IEEE Components and Packaging Technologies* vol. 28, no. 3 (2005), pp. 397–407.

27. P. Dutta. Creep and thermal cycling of continuous fiber reinforced metal-matrix composites. *Key Engineering Materials* vols. 104–107 (1995), pp. 673–690.

28. Y.-L. Shen, S. Suresh. Steady-state creep of metal-ceramic multilayered materials. *Acta Materialia* vol. 44 (1996), pp. 1337–1348.

29. H. J. Frost, M. F. Ashby. *Deformation-Mechanism Maps: The Plasticity and Creep of Metals and Ceramics*. New York: Pergamon Press, 1982.

30. R. Nagarajan, I. Dutta, J. V. Funn, et al. Role of interfacial sliding on the longitudinal creep response of continuous fiber reinforced metal-matrix composites. *Materials Science and Engineering A* vol. 259, no. 2 (1999), pp. 237–252.

31. K. A. Peterson, I. Dutta, M. W. Chen. Diffusionally accommodated interfacial sliding in metal-silicon systems. *Acta Materialia* vol. 51, no. 10 (2003), pp. 2831–2846.

32. J. V. Funn, I. Dutta. Creep behavior of interfaces in fiber reinforced metal–matrix composites. *Acta Materialia* vol. 47, no. 1 (1999), pp. 149–164.

33. P. Kumar, I. Dutta. Influence of electric current on diffusionally accommodated sliding at hetero-interfaces. *Acta Materialia* vol. 59, no. 5 (2011), pp. 2096–2108.

34. P. Kumar, Z. Huang, I. Dutta, et al. Fracture of Sn-Ag-Cu solder joints on Cu substrates. I. Effects of loading and processing conditions. *Journal of Electronic Materials* vol. 41, no. 2 (2012), pp. 375–389.

35. F. X. Che, W. N. Putra, A. Heryanto, et al. Study on Cu protrusion of through-silicon via. *IEEE Transactions on Components, Packaging and Manufacturing Technology* vol. 3, no. 5 (2013), pp. 732–739.

36. F. Dalla Torre, P. Spatig, R. Schaublin, M. Victoria. Deformation behaviour and microstructure of nanocrystalline electrodeposited and high pressure torsioned nickel. *Acta Materialia* vol. 53, no. 8 (2005), pp. 2337–2349.

37. A. Heryanto, W. N. Putra, A. Trigg, S. Gao. Effect of copper TSV annealing on via protrusion for TSV wafer fabrication. *Journal of Electronic Materials* vol. 41, no. 9 (2012), pp. 2533–2542.

38. C. Okoro, J. W. Lau, F. Golshany. Influence of annealing conditions on the mechanical and microstructural behaviour of electroplated Cu-TSV. *IEEE Journal of Micromechanics and Microengineering* vol. 20, no. 4 (2014), pp. 45032–45037.

39. P. G. Shewmon. *Diffusion in Solids*. New York: McGraw-Hill, 1963.

40. X. Liu, Q. Chen, V. Sundaram, R. R. Tummala, S. K. Sitaraman. Failure analysis of through-silicon vias in free-standing wafer under thermal-shock test. *Microelectronics Reliability* vol. 53 (2013), pp. 70–78.

Index